From Blasting to Grinding: The Different Stages of Ore Breakage in Comminution

Jacks

Table of contents

1 Introduction

Comminution is an essential part of mineral processing which involves the breakage of ore to achieve size reduction. Its aim is to increase the likelihood of separation and liberation of valuable minerals from gangue in subsequent processing stages (Napier-Munn et al., 1996; Fuerstenau, 2003; Wills and Napier-Munn, 2015). Ore breakage in comminution occurs in three forms, namely; blasting, crushing and grinding (Kapur et al., 1997; Le Pham, 2011; Yahyaei et al., 2016). During blasting, ore breakage occurs by means of explosive devices. In crushing, it occurs via mechanical compression of ore against hard surfaces (Unland and Szczelina, 2004). In the case of grinding, ore breakage takes places through a combination of two or more of: impact (ores drop against a rigid body or media drop against ores), abrasion (ores of similar size shear against each other) and attrition (relatively big ores grind smaller materials) (Austin, 2002; Le Pham, 2011; Wills and Napier-Munn, 2015).

Comminution is well known to be highly energy-intensive. Previous studies have shown that comminution typically accounts for 50% of the energy consumption in a mineral processing circuit, and globally comminution activity accounts for about 0.2% of the world's electricity supply (Napier-Munn et al., 1999; Ballantyne and Powell, 2014). In addition, the process has been shown to be inefficient, in that a small fraction of the energy supplied is directly utilised for rock breakage (Tromas, 2008). The remaining energy is either lost as heat (Radziszewski, 2013), carbon dioxide (CO_2) emissions (Norgate and Haque, 2010) or absorbed by the comminution equipment causing wear and degradation (Weerasekara et al., 2013).

Due to the increased scrutiny around energy consumption in industrial activity and concerns over the sustainability of current practices, there is a need to study the common approaches to comminution with a view to optimising the process. The understanding of fundamental mechanisms in comminution, such as impact breakage of single rock specimen, has been identified as the key to improving energy efficiency (Napier-Munn, et al., 1996; Tavares and King, 2004; Bonfils, 2017).

In order to examine the fundamentals of rock breakage, several laboratory-scale breakage devices have been developed. These include the drop weight tester (Genc et al., 2004), twin pendulum tester (Salman et al., 2007) and rotary breakage tester (Kojovic et al., 2010). These devices have mainly been used to obtain empirical correlations relating input energy to product size, with a view to predicting the extent and rate of breakage in industrial-scale machines (Narayanan and Whiten, 1988; Verret et al., 2011). Other devices such as split Hopkinson pressure bars (Bbosa, 2007) and the impact load cell (Bourgeois and Banini, 2002), in the form of the ultra-fast load cell (UFLC) or short impact load cell (SILC), have been utilised to provide quantitative ore property measurements based on rock mechanics, such as strength, stiffness and fracture energy (Tavares, 1999; Bonfils, 2017).

Several factors have been shown to affect breakage dynamics and the behaviour of rock specimens during breakage tests. These include mass and impact velocity of media, pre-existing cracks, rock shape and mineralogical composition (Sikong et al., 1990; Schönert, 1991; Tavares and King, 1998; Chandramohan et al., 2010; Shi and Kojovic, 2007; Salman, 2007; Dube, 2016). Consequently, exploring the role of each individual factor in causing variabilities in breakage test results remains a challenge. Current approaches to breakage testing customarily make use of real rock specimens, which have inherent flaws (pre-existing cracks), irregular shapes and little-known mineralogical composition, such that results are often difficult to reproduce.

However, an alternative approach, which uses synthetic rocks for breakage experiments has emerged (Squelch, 2018; Barbosa et al., 2019). A synthetic rock can be described as an artificial rock specimen in which the materials, compositions and the process of manufacturing are known and controlled (Johnston and Choi, 1986). It has the advantage of being able to reproduce the desired shape and size of a rock specimen (Gell et al., 2019). It is often used as an alternative to real rocks where heterogeneities may influence observed behaviour during standard rock strength characterisations (Lee et al., 2011; Mei et al., 2017; Nong and Towhata, 2017). A recent study by Barbosa et al. (2019) has demonstrated its applicability in comminution experiments conducted with a SILC device. While the use of synthetic rock specimens on breakage devices is valuable, particularly in providing standard material strength characterisations, it is nevertheless impractical in examining micro-scale detail such as the rock specimen's internal stress propagation, crack propagation and isolating the effect of a single factor from standard experiments.

Experimental studies alone have been found to be limited in quantifying the associated energy and underlying mechanisms of rock breakage (Delaney et al., 2015). Limitations also exist in the number of experiments that can be plausibly investigated whilst varying different parameters and the related cost in conducting such studies (Charikinya, 2015; Montgomery,

2017). Numerical techniques have emerged as a means of complementing laboratory studies, acting as a "virtual laboratory" through which minute details that are impractical to characterise via experiments can be analysed. Numerical studies also enable specific parameters to be altered under carefully chosen virtual conditions and the examination of their effects in isolation on the types of macro-scale responses measured by experiments (Han et al., 2017).

Discrete Element Method (DEM) is a numerical technique based on Newton's law of mechanics, initially designed to study the flow of granular materials (Cundall and Strack, 1979). The interaction between these materials and predefined environment are simulated using contact models which calculates the forces and associated energy loss (Cundall and Strack, 1979). As DEM has been refined, it has demonstrated the capability to study dynamic scenarios such as; earthquake, rock fracture and comminution (Mora et al., 1993; Potyondy et al., 1996; Morrison et al., 2007). These scenarios necessitated constructing numerical rock specimens of desired shape and size and the implementation of breakage models within DEM. A number of breakage models with variation in mathematical formulations and their mode of implementation have been utilised in comminution studies. The most prominent are the Discrete Grain Breakage (DGB) method (Potapov and Campbell, 1994), Particle Replacement Method (PRM) (Cleary, 2001) and the Bonded Particle Model (BPM) (Potyondy and Cundall, 2004). The DGB and PRM models are semi-empirical; replacing individual discrete rocks with a population of smaller rocks in accordance with a prescribed appearance function. In BPM, a rock specimen is represented by connecting/contacting discrete entities. The BPM aims to simulate rock breakage with fragmented sizes and shapes determined dynamically i.e., shapes and size distributions of fragments arise naturally.

In essence, ore breakage in comminution circuits occurs via several breakage mechanisms. Impact breakage of single rock specimen is often employed to understand energy usage and fundamentals of the breakage process. The SILC device has been demonstrated as a viable testing device to study energy utilisation and factors that influence breakage behaviour during impact loading on real rock specimens (ore). In order to simplify the variabilities inherent due to ore heterogeneity, synthetic rock specimens have been used as a basis to study the load response and fracture. Even with this simplification, stress distribution and crack propagation within a rock specimen, as well as how these are affected by influencing factors are impractical to investigate experimentally in isolation. DEM offers a promising avenue to overcoming these limitations.

Arising from the background, it is inferred that a synthetic rock specimen is similar in some respects to a DEM rock specimen, in that rock specimens of particular shapes and sizes can be synthesised to give desired attributes. These specimens can be created to be homogeneous i.e., without variations due to pre-existing cracks or mineralogical compositions. Thus, DEM can be used to mimic a synthetic rock specimen of specified material properties, and thereafter to match the mechanical properties through a rigorous calibration. Furthermore, the load response on a rock specimen from a numerical SILC simulation can be extracted and directly compared with the actual experiment. This provides another opportunity to validate the result of DEM simulation. Several studies have been conducted using rock breakage tests to investigate effects such as media mass and velocity (Dube, 2016; Edwards, 2016) as well as the effect of rock shape on rock breakage (Unland and Al-Khasawneh, 2009; Chandramohan et al., 2010; Barbosa et al., 2019). While some experimental success on investigating the effect of media mass and velocity has been recorded (Dube, 2016; Edwards, 2016) as well as the effect of particle shape on rock breakage (Unland and Al-Khasawneh, 2009; Chandramohan et al., 2010; Barbosa et al., 2019). However, no known experimental study has been able to isolate and examine factors such as the extent of pre-existing cracks or the degree of heterogeneity mainly due to the impracticality of ensuring that sufficient identical specimens are utilised for breakage tests under a controlled and reproducible environment. Therefore, the primary objectives of this work are: to use DEM to reproduce fracture characteristics of synthetic rocks in SILC experiments by DEM simulations and thereafter to use this numerical methodology to investigate in isolation the effect of pre-weakening based on the extent of pre-existing cracks and change in breakage behaviour due to mineralogical composition. To achieve these objectives, a numerical methodology for single rock breakage by an impact is developed. The BPM in DEM is employed to simulate the fracture characteristics of rock specimen breakage with results compared using conventional load parameters and measured fracture force patterns from experiments.

1.1 Hypothesis

The following hypotheses form the basis for this study:

- The BPM in DEM can be used to calibrate mechanical properties and conduct breakage simulations in line with standard tests. This is because the macroscopic breakage behaviour is controlled by the microstructural model parameters i.e., model parameters are related to macroscopic mechanical properties. The BPM in DEM can demonstrate that varying the model parameters will result in changes in the macroscopic mechanical properties during standard tests.

- The BPM in DEM can examine in isolation the effect of pre-weakening and mineralogical composition. This is because:

  i. bonds between adjacent discrete entities within a homogeneous rock specimen can be removed to mimic cracks. The BPM in DEM can demonstrate that increasing the degree of pre-weakening or crack density (i.e., increase in discontinuities between connecting discrete entities) will result in lower fracture force during impact loading in SILC breakage tests.

  ii. bonded entities within a homogeneous rock specimen can be grouped and ascribed with desired mechanical properties to mimic a heterogeneous rock specimen (rock with mineral phases). The BPM in DEM can demonstrate that for an ore composed of two-phase minerals (binary ore), fracture predominantly occurs around the weaker phase. It can further demonstrate that increasing the composition of the weaker phase leads to a lower fracture force as the binary rock specimen creates a higher crack propagation network about the weaker mineral.

1.2 Objectives and scope

This work primarily focuses on the impact breakage of single rock specimens in a SILC device using the BPM in DEM. It initially evaluates the robustness of the BPM for the numerical methodology employed herein and thereafter extends the model to conduct impact breakage tests as a "virtual SILC device". As a first attempt, this study is restricted to a homogeneous composition of rock specimens and simplified heterogeneous rock specimens, i.e. homogeneous specimen with either pre-existing cracks or binary (two-phase) mineral composition.

Therefore, the main objectives of this research are:

- **To calibrate the BPM parameters against measurable macroscopic mechanical properties and develop calibration relationships.**
  The purpose of the calibration herein is to develop a methodology which is applicable to a range of comminution systems. Here, uniaxial compression and direct tension simulation tests will be used to measure the macro-mechanical properties in accordance with standard geotechnical testing practice (ISRM standards). Model parameters will be systematically varied against macroscopic mechanical properties. The measured parameters will then be fitted into a fitting function to describe a relationship between the model parameter and macroscopic properties.
- **To generate data for impact breakage study in a SILC device using DEM.**
  The impact force, free surface generated as well as stress data at discrete time-steps will be recorded and examined according to conventional experimental measures. A case study of a simulated breakage behaviour of a single rock specimen will be considered to demonstrate the analysis of data from the DEM simulation. Homogeneous DEM-synthetic rock specimens will be used to conduct "virtual SILC" simulations under a closely controlled environment.

- **To develop a robust BPM-DEM methodology to measure fracture characteristics.**
  The robustness of the model will be evaluated by considering the effect of model resolution, sample size and variation of macroscopic mechanical properties as criteria to relate against standard material properties.

- **To compare the impact breakage of single rock specimens using SILC experiment and numerical simulations.**
  SILC breakage data using synthetic rock specimens will be gathered and analysed. The experimental work conducted by Barbosa et al. (2019) will be used to compare against numerical simulations as a basis to validate the methodology.

- **To investigate factors that influence rock breakage.**

The extent of pre-weakening and mineralogical composition will be varied to investigate their effects in isolation on breakage behaviour and standard experimental measurements.

Fig. 1.1 illustrates a simplified schematic of the scope of this work in relation to the highlighted objectives. A chart detailing the approach to the current methodology.

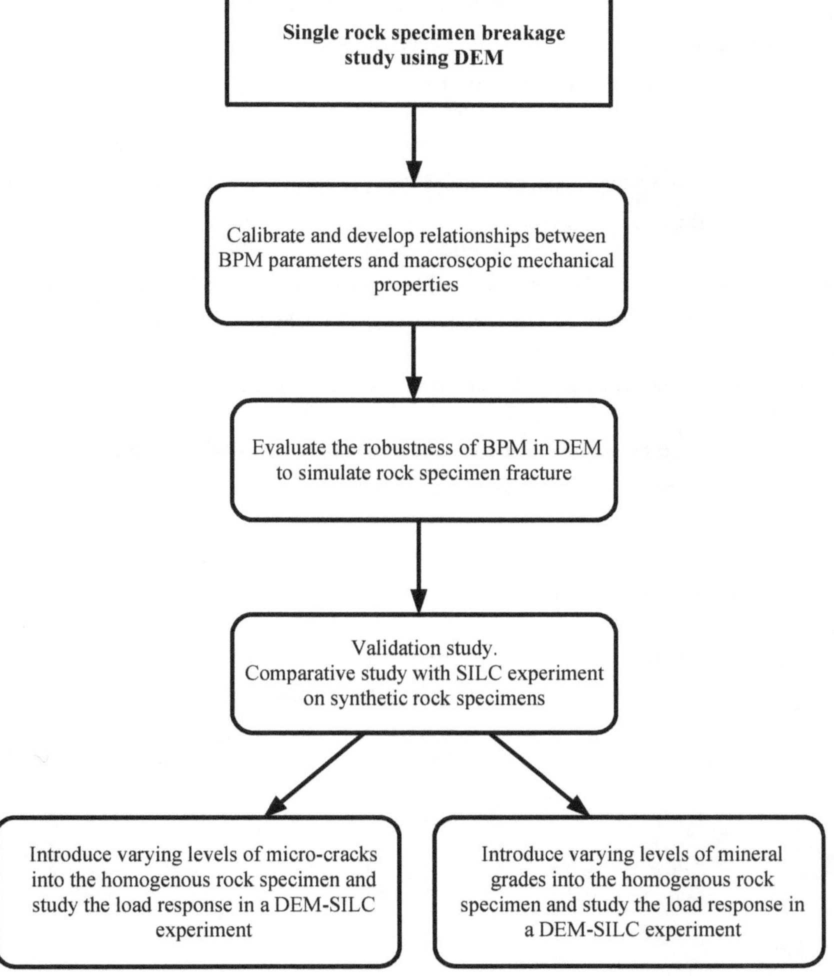

Fig. 1.1: Figure showing the scope of the current work

1.3   Plan of development

The structure of this book is as follows:

- Chapter 2 presents the literature review. This comprises of a review of rock breakage in comminution and the underlying theory of fracture and rock mechanics. It also includes a brief revision of comminution testing devices (with emphasis given to the SILC), as well as the fundamentals DEM and the bonded particle breakage model, employed. A summary of some of the influencing factors identified to affect rock breakage in comminution is also covered.

- Chapter 3 describes the calibration approach and the development of the empirical calibration relationships between BPM parameters and macroscopic mechanical properties.

- Chapter 4 details the numerical methodology employed, and approach selected for data analysis as well as a case study of a simulated breakage behaviour.

- Chapter 5 discusses an evaluation of the BPM used in this work.

- Chapter 6 presents a comparative study of the current work with SILC experiments using synthetic rocks of different shapes and sizes.

- Chapter 7 reports the results of the load response on the extent of pre-weakening and mineralogical composition.

- Chapter 8 provides a general discussion on the key outcomes of the investigation and highlights the novel contribution made to the existing research.

- Chapter 9 summarises the conclusions and poses recommendations for future work.

2 Literature review

2.1 Rock breakage in comminution

Comminution is well known to be an essential part of mineral processing (Wills and Napier-Munn, 2015). It primarily aims to break ore into smaller sizes to increase the likelihood of liberation of valuable minerals from gangue and to enhance the efficiency of the down-stream processes (Napier-Munn et al., 1996). Comminution is a highly energy-intensive and inefficient process. Previous reports have stated that it accounts for 50% of the energy consumption on a mineral processing circuit, while it is estimated that only 3% of this energy is directly utilised in ore breakage (Napier-Munn et al., 1999; Tromas, 2008; Ballantyne and Powell, 2014).

A fundamental understanding of rock breakage mechanisms has been identified as the likeliest means of making significant improvements to energy efficiency in comminution circuits (Napier-Munn, et al., 1996; Tavares and King, 2004; Bonfils, 2017). Rock breakage in this context can be defined as the reduction in the size of a large rock specimen into a targeted size under an applied stress condition (Tavares and King, 1998; Herbst et al., 2003). In comminution circuits, rock breakage occurs in three basic forms; blasting, crushing and grinding. Blasting utilises explosive devices on the natural bed of rocks or in-situ to achieve size reduction for easy transportation of rock fragments (Yahyaei et al., 2016). Size reduction during crushing and grinding is achieved through mechanical means with crushing typically involving large reduction ratios while grinding involves the breakdown of relatively fine materials (Kapur et al., 1997).

Rock breakage is influenced by the inherent ore properties and the mode of stressing in the comminution equipment (Potapov and Campbell, 2001; Tavares, 2007). Four main mechanisms of rock breakage have been identified: abrasion, attrition, compression and impact. Schematics of these are given in Fig. 2.1a to d respectively (Hogg, 1999; Austin, 2002). Abrasion occurs by virtue of shearing rock specimens of similar sizes against each other or grinding media leading to a gradual degradation of the rock specimens (Hogg, 1999). Attrition occurs when smaller rock specimens are ground by larger ones (King, 2001). During compression, rock specimens are squeezed against hard surfaces leading to fracture under partial or full confinement (Unland and Szczelina, 2004). Impact breakage is when a rock specimen drops against a rigid body or grinding media drops against rock specimen causing the rock specimens to fracture (Austin, 2002). Among these mechanisms, comminution by impact is the most widely used basis through which rock specimen breakage has been studied.

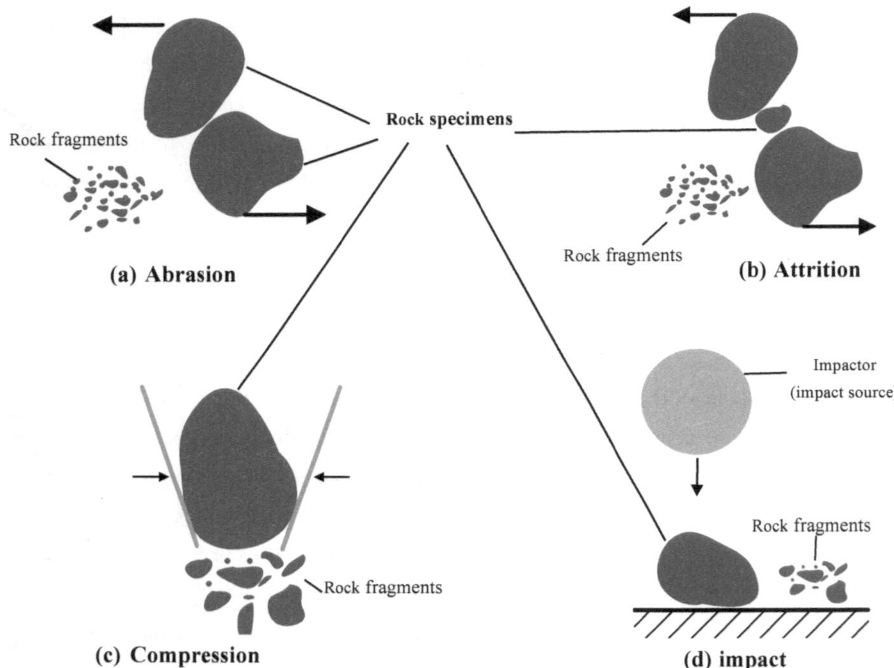

Fig. 2.1: Rock breakage mechanisms (Modified from Chikochi, 2017; Napier-Munn et al., 1996)

All the breakage mechanisms highlighted in Fig. 2.1 occur fundamentally through the application of stress on rock specimens. Stress is applied on rock specimens through any of the following three modes:

- **Shearing:** opposing forces parallel to each other act on the rock specimen. This application is found during abrasion and attrition (Figs 2.1a and b)

- **Compressive stress:** the length of rock specimen is reduced by confinement which leads to fracture under compression (Fig. 2.1c)

- **Tensile stress:** rock specimen is elongated by the action of the applied load. This leads to fracture via stress under tension (Fig. 2.1d).

The behaviour of rock specimens under an applied load has mainly been studied from two perspectives; macroscopic (rock mechanics) and microscopic (fracture mechanics) viewpoints (Napier-Munn et al., 1996).

2.1.1   Rock mechanics

Rock mechanics is the study of the deformation of a rock specimen until the point of fracture. Several laboratory experiments have been devised to study fracture from this standpoint (Brace, 1961; Chaboche, 1988; Goodman, 1989; Jaeger, 2009), which have mainly been conducted using uniaxial compressive/tensile loading tests on rock specimens. A stress-strain curve typically used to explain the deformation behaviour in such tests is shown in Fig. 2.2. The key features and points that the deformation path follows are shown in Table 2.1 (Nesetova and Latjai, 1973; Eberhardt, 1998; Hudson and Harrison, 2000; Liu et al., 2007).

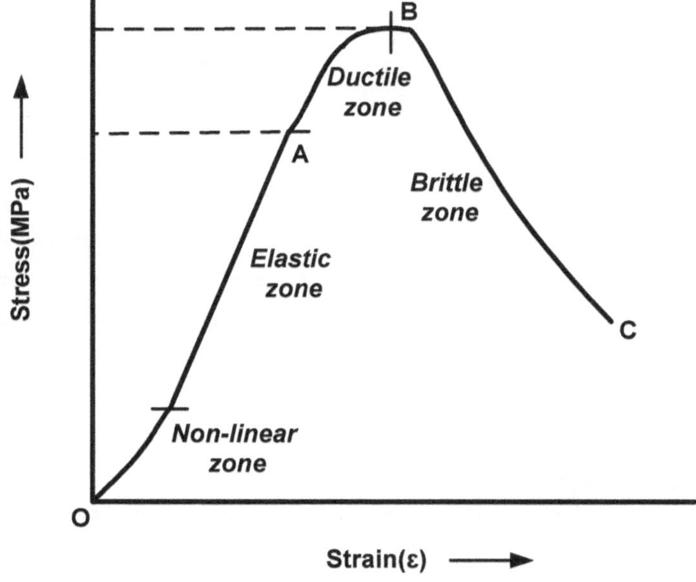

Fig. 2.2: Typical stress-strain curved obtained during uniaxial compressive loading. Adapted from Napier-Munn et al. (1996)

Table 2.1: A summary of key points during uniaxial compressive loading test

Zone	Key point	Keynote(s)
Non-linear zone	Crack closure	Occurs at the initial loading state where existing cracks are re-orientated.
Elastic zone	Linear elastic deformation	Major existing cracks are closed and the slope of the axial stress vs axial strain graph at this zone is the elastic modulus.
Ductile zone	Elastic limit	Stress level approaches the maximum and non-reversible damage occurs.
Brittle zone	Critical energy release	Larger cracks begin to form as the energy is released also known as the point of ultimate strength.

2.1.2 Fracture mechanics

Fracture mechanics is the field of study that focuses on the loading condition of a load-bearing solid body that results in failure due to the enlargement of a major crack present in the body (Kanninen and Poplar, 1985). Failure can be defined as the point at which a rock specimen can no longer bear the load, breakage usually occurs under tensile loading to produce coarse fragments (Le Pham, 2011). For some materials, this field of study is termed "brittle fracture mechanics" because breakage, for instance in comminution and some engineering fields, primarily occurs through brittle fracturing. Griffith (1921), the pioneer of this approach, examined an arbitrary body made up of atoms that are held together by brittle bonds with a discontinuity i.e., a crack within the body. He postulated that brittle fracture in an elastic material initiates through tensile stress (σ_t) concentrations at the micro-crack tips that extend until the point of failure as shown in Fig.2.3.

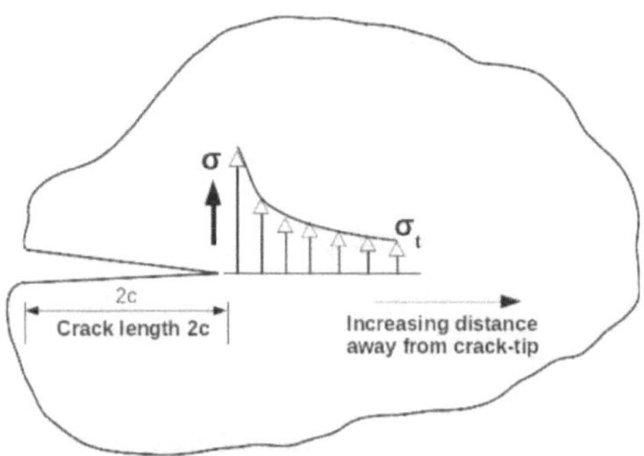

Fig.2.3: An illustration of fracture mechanics according to Griffith (1921). Elastic tensile failure stress (σ_t) at a crack tip in a body. Re-drawn from LePham (2011).

Griffith (1920) determined the energy required for propagation and extension of such a crack and deduced that the energy required to go from one state of "un-brokenness" to a state of "brokenness" occurred via a given length which is termed Griffith's crack length (Griffith, 1920). According to this hypothesis, the total potential energy (U_{total}) as a result of the crack extension is equal to the elastic strain energy (U_e) stored in the body and the fracture surface energy (U_s) (Eq.2.1).

$$U_{total} = U_e + U_s \qquad\qquad 2.1$$

The criterion used in brittle fracture theory to predict fracture behaviour is also given by the energy balance proposed by Griffith. A state of equilibrium is attained when an increase in crack length (2c) illustrated in Fig.2.3, does not produce a change in the total energy to the system at a constant applied stress (σ) (Eq. 2.2). This implies that an increase in the fracture surface energy will be balanced by a decrease in the elastic strain energy (Eq. 2.3).

$$\frac{dU_{total}}{dc} = 0 \qquad\qquad 2.2$$

$$\frac{dU_s}{dc} = -\frac{dU_e}{dc} \qquad\qquad 2.3$$

Inglis (1913) solved for the above case by considering a plate with the presence of a crack and negligible thickness subjected to tensile loading and derived Eqs. 2.4 and 2.5.

$$U_e = -\frac{\pi^2 c^2 \sigma^2}{Y} \qquad\qquad 2.4$$

$$U_s = 4c\gamma \qquad\qquad 2.5$$

Where c is the half of the crack length, Y is Young's modulus and γ is the atomic bond. Therefore, Eq. 2.5 becomes:

$$\sigma\sqrt{\pi c} = \sqrt{2Y\gamma} \qquad\qquad 2.6.$$

For an ideal brittle material, fracture initiation occurs when the left-hand form ($\sigma\sqrt{\pi c}$) reaches a critical value known as the critical value stress intensity factor or fracture toughness (K_f).

Therefore, the fracture stress (σ_f) for an ideal brittle material is expressed as:

$$\sigma_f = \sqrt{\frac{2Y\gamma}{\pi c}} \qquad\qquad 2.7$$

Tavares (1997) reported a review of the modification of Eq.2.7 by Irwin (1947) and Orowan (1949) to account for the specific fracture energy or fracture surface energy (γ_f) as shown in Eq. 2.8

$$\sigma_f = \sqrt{\frac{2Y\gamma_f}{\pi c}} \qquad\qquad 2.8$$

Irwin (1956) further defined the energy release rate (G) for an increase in the extension of a crack in a thin plate as:

$$G = -\frac{dU_e}{dc} = \frac{\sigma^2 \pi c}{Y} \qquad\qquad 2.9$$

The relationship associating the fracture toughness, fracture surface energy and the critical strain energy release rate (Gc) was then as expressed in Eq. 2.10 (Tavares, 1998)

$$K_f = \sqrt{YG_c} = \sqrt{2\gamma_f Y} \qquad\qquad 2.10$$

This relationship has been a significant contribution to the field of fracture mechanics and related structural design fields (Tavares, 1998).

Apart from the Griffith failure criterion highlighted above, the other two most commonly known and quoted methods are the Hoek-Brown failure (Eq.2.11) and Mohr-Coulomb failure (Eq.2.12) criteria that can be used to quantify micromechanical parameters of a rock specimen (Sjöberg, 1997).

$$\sigma_1 = \sigma_3 + \sqrt{m\sigma_c + \sigma_3 + s\sigma_c^2} \qquad\qquad 2.11$$

$$\tau = C + \sigma_n tan\varphi \qquad\qquad\qquad\qquad\qquad\qquad 2.12$$

Where: m and s are constants that depends on rock mass

σ_c = UCS of unbroken rock specimen

σ_1 = major principal stress

σ_3 = minor principal stress

τ = shear strength

σ_n = normal stress

C = cohesion (cohesive strength) of rock specimen

φ = internal angle of friction of rock specimen

Amongst these criteria, the Mohr-Coulomb failure criterion is the simplest to implement due to the linearity of the shear criterion and characterisation by only two parameters; Cohesion (C) and the internal angle of friction (θ) (Labuz and Zang, 2012)

2.2 Rock breakage characterisation methods

Rock breakage characterisation methods span across the study of rock and fracture mechanics as well as mineral processing. The standards of rock breakage characterisation that have been established in these fields can be grouped into three (Napier-Munn, et al., 1996):

- Rock and fracture mechanics technique

- Mineral processor's approach

- Single rock specimen breakage tests

These techniques have different applications in the various fields and are described briefly in the following sub-sections.

2.2.1 Rock and fracture mechanics technique

This method relies on both compressive and tensile load on rock specimen to measure macroscopic properties of rock. During compressive loading, the load response on rock specimen obtained from the stress-strain curve is used to determine the macroscopic Young's modulus, Poisson's ratio and unconfined compressive strength (UCS) of rock (Pharr et al., 1992). The tensile load applied to rock specimen, popularly known as the standard Brazilian test, is used to determine the tensile strength of rock specimen as well the mode of failure from crack initiation to propagation (Rocco et al. 1999).

2.2.2   Mineral processor's approach

This approach is otherwise known as the "standard grindability test". This method is adopted by mineral processing engineers to estimate the energy requirement of the breakage process to achieve a specific product size. The most common example is the Bond model (Eq. 2.13) that has been found to work well for the ball and rod mill for which it is still widely used (Napier-Munn et al., 1996).

$$W=10Wi\left(\frac{1}{\sqrt{P_{80}}} - \frac{1}{\sqrt{F_{80}}}\right)$$ 2.13

W = work input (kWh/t)

Wi = Work index which is rock specimen specific constant (kWh/t)

P_{80} = 80% passing size of the product (μm)

F_{80} = 80% passing size of the feed (μm)

According to Napier-Munn et al. (1996), there are ranges of values of Wi that corresponds to ranges of vales UCS to categorise the property of such rock (Table 2.2)

Table 2.2: Relationship between uniaxial compressive strength, UCS, and Work index (Napier-Munn et al., 1996)

Rock category	Soft	Medium	Hard	Very hard
UCS (MPa)	50-100	100-150	150-250	>250
Wi (kWh/t)	7-9	9-14	14-20	>20

2.2.3   Single rock specimen breakage test

The single rock specimen breakage test can be seen as an interwoven technique between the rock and fracture mechanics technique and the mineral processor's standard grindability test. It is regarded as the best technique through which fundamental breakage of rock specimen is well-understood (Tavares, 2004). This test has been used in various studies to unravel a number of fundamental aspects of comminution researches such as:

- Energy usage and energy loss during comminution (Tavares, 1999)

- Relationship between input energy to the product size (Vogel and Peukert, 2004)

- Material response to deformation under an applied load (Tavares and King, 2004)

- Determination of rock specimen breakage parameters for rock characterisation, modelling and simulation (Napier Munn et al., 1996)

In order to examine these fundamentals of rock specimen breakage, several laboratory-scale breakage devices have been developed to act as a proxy for examining the basic principles of comminution.

2.3   Laboratory scale breakage devices

Several laboratory scale breakage devices have been developed to study and interpret rock breakage (Weichert and Herbst, 1986; Briggs and Bearman, 1996; Fandrich et al., 1998; Bourgeois and Banini, 2002; Kojovic et al., 2008). These devices include; the twin pendulum tester, rotary breakage tester (RBT), drop weight tester (DWT), split Hopkinson pressure bars (SHPB) and impact load cell. A summary of the main strengths and weaknesses of these devices are given in Table 2.3

Table 2.3: A summary of the advantages and disadvantages of the various laboratory-scale breakage testing devices

Comminution testing device	Advantage(s)	Disadvantage(s)	Reference(s)
Pendulum tester	Straight forward to operate	- Time-consuming to set up - Limited size range	Salman et al. (2007)
Drop weight tester	- Relatively wide range of input energy - Suitable for single rock specimen and bed breakage	Difficult to measure the exact amount of energy used for rock breakage	Napier-Munn et al. (1996); Bearman et al. (1997); Tavares (1999)

	• Suitable for repeated drop test		
Rotary breakage tester	• Greater precision of the measured input energy • Can test a relatively high number of rock specimens • Incremental breakage	• The amount of input energy utilised by rock breakage is not quantified • No bed breakage • May result in multiple impact	Tavares (2007), Kojovic et al. (2010)
Split-Hopkinson bar	• Detailed representation of breakage event • Measures energy associated with impact breakage of rock specimen.	• The operational procedure is time-consuming • No bed breakage	Bbosa et al. (2006)
Impact load cells	• Simplified methodology relative to Split-Hopkinson bar • Measures the amount of energy used for rock breakage • Bed breakage	A relatively low range of input energy	Bourgeois and Banini (2002)

The load cell, which has the attributes most related to the scope of this work, will be discussed in detail. Since this device is a hybrid of the drop weight tester and Split-Hopkinson bar, these are first examined.

2.3.1 Drop weight tester

For the drop weight tester (schematic given in Fig. 2.4), the impactor (drop weight) falls under restriction at guide rails which offer minimal resistance along the vertical direction which impacts directly on the rock specimen placed on a stationary anvil or die (Genc et al., 2004; Eksi et al., 2011).

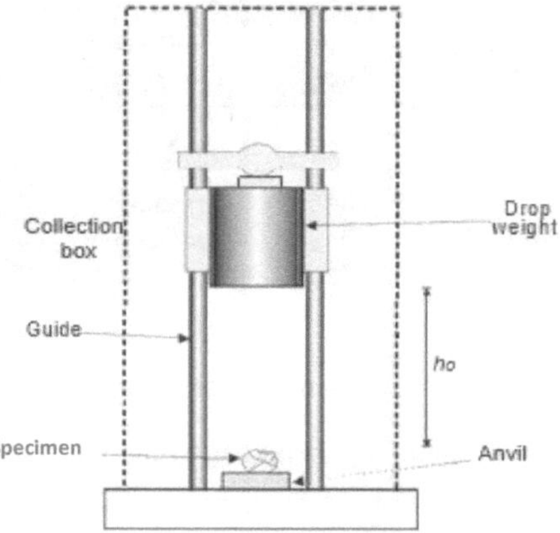

Fig. 2.4. Schematic diagram of a drop weight tester. Adapted from (Tavares, 2007)

The input energy is defined as the potential energy at a set drop height before impact. (Genc et al., 2004; Eksi et al., 2011). It is typically assumed that all the potential energy is delivered into the rock specimen for breakage (Napier-Munn et al., 1996; Genc et al., 2004). However, Radziszewski and Laplante (2006) reported energy losses due to friction of the guide rails and rebound of the drop weight.

2.3.2   Split-Hopkinson bar

The split Hopkinson bar device (schematic given in Fig. 2.5) consists of three cylindrical steel rods named; the striker, incident bar and transmitter bar. The rock specimen is placed in-between the incident bar and the transmitter bar. The striker is fired by some of launcher and impacts against the incident bar. This sets up a strain wave which travels through the incident bar, breaking the specimen and partially transmitting stress to the second bar mounted with strain gauges (Gray, 2000; Tavares, 2007). The strain gauges measure the strain and data generated are recorded on computer to determine the impact response of the sample.

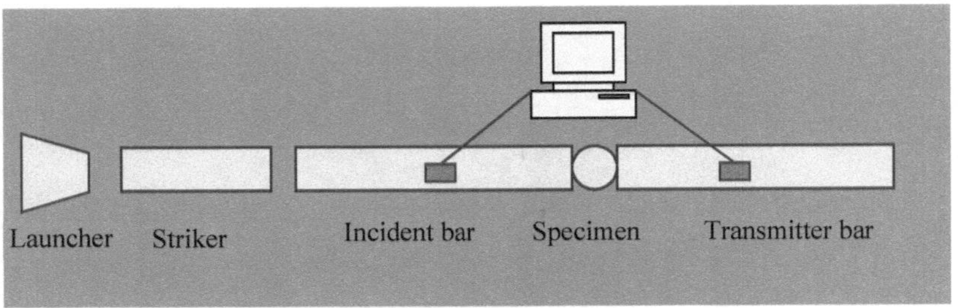

Fig. 2.5: Schematic diagram of a split-Hopkinson bar. Adapted from Bbosa (2007)

The main benefit of using the split Hopkinson bar is that the input energy, as well as the energy absorbed by the materials, can be calculated. The device also provides a detailed characterisation of the impact event. This information can be interpreted and used to study the interaction between stress and energy that ultimately leads to breakage (Bbosa, 2007). However, the major constraints of the device are the limitations of the size range of rock specimens and the time-consuming nature of each test. It is only suitable for the breakage of single rock specimens and the process occurs at high strain rates (Bbosa, 2007). This is distinct from the comminution environment which involves breakage at considerably lower strain rates (Saeidi et al.,2017).

2.3.3   Impact load cells

The impact load cells (ILC) was developed from an adaptation of the drop weight tester and split- Hopkinson bar (Weichert and Herbst, 1986; Tavares and King, 1998). This device operates using a similar principle to the drop weight tester. Instead of drop weight, a steel ball under freefall acts as the impactor. As a substitute to an anvil, a vertical cylindrical rod mounted with strain gauges is used such that the longitudinal stress can be obtained in a similar manner to the split-Hopkinson pressure bar. Fig. 2.6 shows a schematic diagram of an ILC device and the main components.

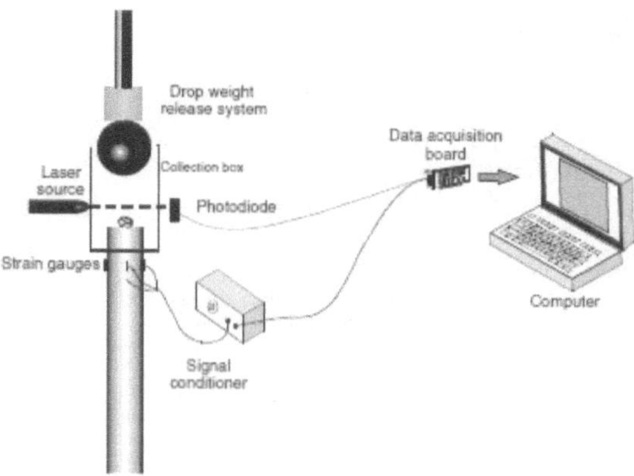

Fig. 2.6: Schematic diagram of an ILC. Adapted from Tavares (2007)

For a breakage test on an ILC, the freefalling steel ball passes through a laser beam which triggers a digital oscilloscope to begin the recording of the impact event (Dube, 2016). The impact of the steel ball on the rock specimen causes a compressive wave to pass through the steel rod which creates resistance change while passing through the strain gauges (Bourgeois and Banini, 2002; Dube, 2016). This change in resistance produces a change in voltage that is recorded over time and displayed on the computer screen as a voltage-time graph (King and Bourgois, 1993; Tavares; 1998; Bourgeois and Banini, 2002). The captured data of this graph are further processed to determine the force versus time history of the impact event (Tavares and King, 2004). Typical force-time histories of impact events on different rock specimens obtained from an ILC device is shown in Fig. 2.7

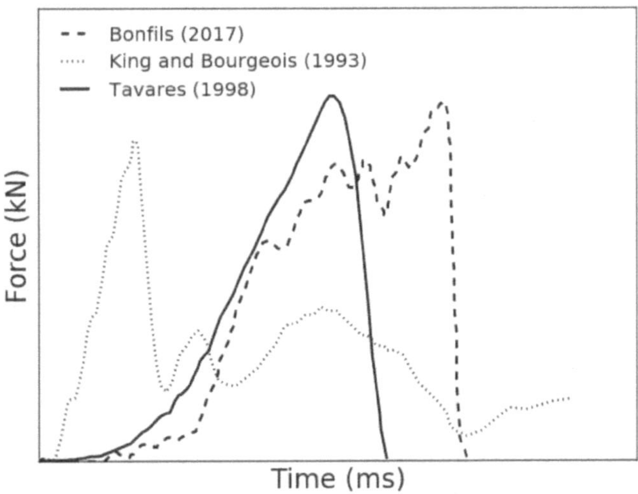

Fig. 2.7: An Illustration of the force-time history of different rock specimens obtained from impact load cells. (King and Bourgois, 1993; Tavares; 1998; Bonfils, 2017)

The main benefits of the ILC are; accurate measurement of the force-time relationship, suitability for single rock specimen and bed of rocks. The device, however, is only suitable at a relatively low range of input energy (Bourgeois and Banini, 2002). There are two variations of the impact load cell device; the ultra-fast load cell (UFLC) and the short impact load cell (SILC) (Bourgeois and Banini, 2002). The SILC is a portable version of UFLC. The rod length of SILC is about 25% of the rod length of a UFLC (Bourgeois and Banini, 2002).

2.4   Impact breakage studies on factors that influence rock breakage

A number of factors that influence rock specimen breakage and ultimately relate to its associated energy have been highlighted in the literature (Schönert, 1991; Shi and Kojovic, 2007; Salman, 2007; Chandramohan et al., 2010; Dube, 2016). While some of these factors can be viewed as "operational", others can be seen as "inherent" within rock specimens. Examples of operational factors include input energy and comminution equipment. Meanwhile, examples of inherent factors include rock specimen shape, the extent of pre-damage in rocks (pre-existing cracks) and mineralogical composition of rock specimens.

For instance, investigations on operational factors which considered energy input in grinding mill equipment have shown that the energy input can be affected by the velocity and the mass of the grinding media (Morrison and Cleary, 2004; Shi and Kojovic, 2007; Salman, 2007; Dube, 2016).

Sikong et al. (1990) and Schönert (1991), investigated the effect of particle size during breakage and reported that the strength of rock specimen increases with a decrease in the size of rock specimen. This was attributed to the reduction of flaws within rock specimen as the size decreases.

The effect of rock specimen shape and loading direction are somewhat related to each other. This depends on the placement or position of the rock specimen relative to the applied load. Chandramohan et al. (2010) investigated the effect of non-flaky, horizontal-flaky and vertical-flaky rock specimens during impact (Fig. 2.8).

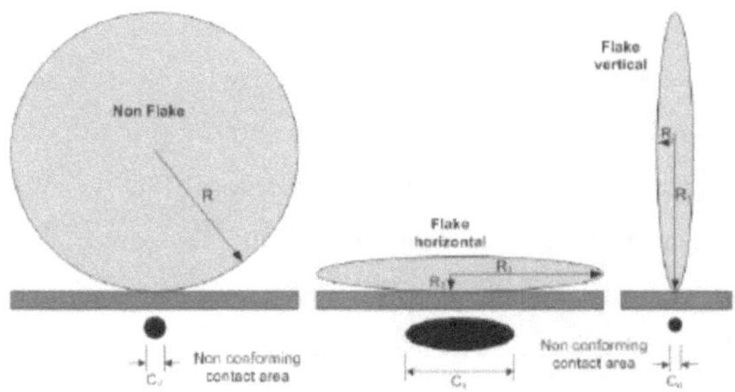

Fig. 2.8: An illustration of particle shapes investigated by Chandramohan et al. (2010)

It was reported that shape does not influence the strength of rock specimen but does have an effect on the hardness which consequently affects the way in which fragments are generated (Chandramohan et al., 2010). In this study the authors presented what they termed as hardness to imply stiffness.

Limited literature exists on the effect of pre-existing cracks and that of mineralogical characteristics on load response. The effect of pre-existing cracks has mainly been investigated in rock mechanics studies to determine the mode of failure (Freund, 1972; Sharon et al.,1995; Han et al., 2015). Mineralogical characteristics of rock specimen on load response have been tested by Tavares and King (1998; 2004) who considered different ores sources (galena, sphalerite, quartz, and corundum). No further information was provided about the actual composition of these ore samples. It was, however, reported that both fracture energy and stiffness varied with the different ore sources.

Due to the complex nature of real rock specimens, exploring the effect of each factor causing variabilities in breakage tests remains a challenge. Current approaches to breakage testing generally make use of real rock specimens with inherent flaws (pre-existing cracks), irregular shapes and little-known mineralogical composition in which results are often difficult to reproduce (Ivars et al., 2011; Gell et al., 2019). However, an alternative technique, which uses synthetic rock specimens has emerged which provides controlled material properties akin to real rocks, with minimal flaws and less heterogeneity (Johnston and Choi, 1986; Ivars et al., 2011; Smith et al., 2014).

2.5 Synthetic rock

A synthetic rock specimen is an artificial rock in which the material make-up and the process of manufacturing are closely controlled. The expected mechanical property of the real rock to be imitated is dependent on the makeup of the material of the synthetic rock (Wong and Chau, 1998). Two main methods are used in the manufacturing of synthetic rock specimens which are: (i) the mould filling and (ii) three-dimensional (3D) printing (Gell et al., 2019). Synthetic rocks produced by mould filling are usually composed of a mixture of a known type of sand, cement or plaster and water, filled in a mould as shown in Fig. 2.9a. The main challenge with this approach has been the challenge of ensuring that the consistency of the mixture matches the desired mechanical properties of the intended rock type (Wong and Chau et al., 1998; Zhou and Zhu, 2018; Gell et al., 2019). 3D printed rock specimens consist of additives and known sand powders printed in layers using a 3D printer (Squelch, 2018), with the printer shown in Fig. 2.9b. This has the advantage of easy repeatability to arrive at a mechanical property close to real rock.

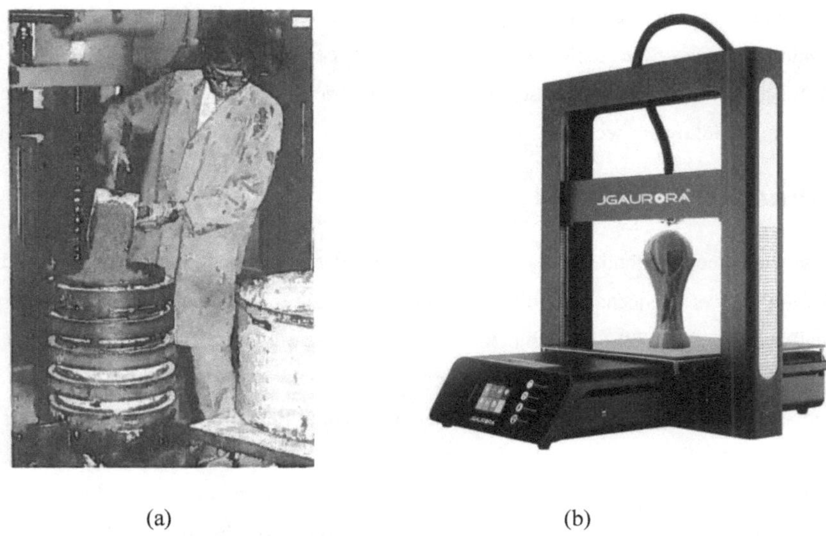

(a)   (b)

Fig. 2.9: (a) A man producing a synthetic rock specimen by mould filling method (Johnston and Choi, 1986) and (b) 3D-printer for printing synthetic rock specimens (Alibaba, 2019)

The use of synthetic rock specimens on breakage devices has been demonstrated to be valuable in providing standard rock strength characterisations (Barbosa et al., 2019). However, the examination of crack propagation, rock specimen's internal stress, and isolating the influence of a single factor are challenging from experiments. Modelling and numerical methods have been proven to be a useful tool to overcome these challenges. The following sections highlight the numerical methods used in comminution studies.

2.6 Modelling rock breakage

Numerical techniques that are being used for breakage studies can be broadly categorised into continuum or discontinuum approaches. For continuum approaches, the phase field method (PFM), performed with finite element method (FEM) is one of the most widely used and is suitable for studying non-uniformity in materials, dynamic interactions and complex boundary conditions (Jing, 2003; Kuhn and Müller, 2010). However, a major limitation of this approach is the mesh sensitivity to discontinuity effects such as the presence of material cracks and sliding under friction and a high computational cost (Mandal et al., 2019; Jing, 2003). Such difficulties can be addressed by adopting discontinuum approaches such as the discrete element method (DEM). DEM has been demonstrated as a viable tool for studying fracture mechanics in many forms including rock blasts, seismic failures and within comminution devices (Weatherley and Ayton, 2012; Weerasekara et al., 2013; Han et al., 2015). Since the current work focuses on DEM, and various aspects of the method will be described in the following sections.

2.7 Discrete element method

The discrete element method (DEM) is a discontinuous numerical modelling approach based on Newton's Laws of mechanics, initially designed to study the flow of granular materials (Cundall and Strack, 1979). In this method, a physical system is represented as an assembly of discrete entities (or elements) that undergo forces due to prescribed interactions with adjacent entities, other entities such as walls, and/or potential fields such as gravity (Matuttis and Chen, 2014). The method explicitly computes the net force acting on each entity at a given time then updates the positions and velocities of the entities by integrating the ensuing equation of motion. A simplified form of such an equation is given in Eq. 2.14 (Cundall and Strack, 1979). In many cases, some forms of artificial damping forces are added to numerically stabilise the ensuing system of equations.

In Eq. 2.14, the instantaneous acceleration of any given entity i is given by the sum of all forces acting on that entity at a prescribed time t. The basic forces are entity-pair interactions or contact force (F_{ij}^{p}), damping (F_{i}^{d}) and gravitation (F_{i}^{g}), as well as wall forces (F_{i}^{w}), although other expressions can be added depending on the scenario to be simulated.

$$m_i \frac{dv_i}{dt} = \sum_{j=1}^{N_i^c} F_{ij}^{p} + F_i^{d} + F_i^{g} + \sum_{w=1}^{N_w} F_i^{w} \qquad 2.14$$

Where:

m_i = mass of entity i respectively at time t. Furthermore, and

v_i = velocities of entity i respectively at time t

N_i^c = the total number of entities in contacts with entity i

N_w is the number of walls within the simulation domain.

The sum of the contact and damping forces is often referred to as the contact model. Several contact models have been developed for DEM, of which the two predominantly used are the linear spring-dashpot model and the non-linear spring-dashpot model (Hertz-Mindlin). Most other contact models tend to be minor modifications of these two methods. The equation of motion (Eq. 2.14) is typically integrated in time via one of the ranges of explicit numerical time integration schemes (Wang et al., 2006; Wang and Mora, 2009). Using one of the simplest such schemes, this integration results in Eqs 2.15 and 2.16:

$$v_i(t + \Delta t) = v_i(t) + \Delta t \frac{dv_i}{dt}(t) \qquad\qquad 2.15$$

$$p_i(t + \Delta t) = p_i(t) + \Delta t\, v_i(t + \Delta t) + \Delta t^2 \frac{dv_i}{dt}(t) \qquad\qquad 2.16$$

Where: p_i is the position of entity i and Δt is the time-step increment. As for any such explicit time integration scheme, a courant criterion governs the maximum permissible time-step increment to ensure numerical stability. For DEM simulations in which entities interact via linear elastic springs (with spring constant K) and entity velocities remain a small fraction of the compressional wave speed, the time-step increment must satisfy Eq 2.17, where m is the mass of the DEM entities.

$$\Delta t \leq \frac{1}{5}\sqrt{\frac{m}{K}} \qquad\qquad 2.17$$

For DEM simulations of this form, DEM entities have a range of radii and spring constants, requiring that m be set equal to the minimum DEM entity mass and K be set equal to the maximum stiffness.

As DEM has been refined, it has proven to be useful for the modelling of a variety of applications including highly dynamic scenarios involving high strain rates (Wang et al., 2000; Griffiths, 2001; Wang et al., 2006). Examples include; earthquake studies and fault propagation (Mora et al., 1993 and Wang et al., 2000), rock fracture (Potyondy et al., 1996; Place et al., 2001; Hentz et al., 2004 and Potyondy and Cundall, 2004) and comminution (Huang et al., 1999; Cleary, 2000 and Morrison et al., 2007). Many such scenarios necessitated the implementation of breakage models within the DEM environment, in addition to contact mechanics appropriate for simulating granular media follow. The following section briefly describes the different classes of DEM breakage models, followed by a detailed account of the method employed in this work.

2.8   Breakage models in DEM

There are three main breakage models employed in DEM investigations (Jiménez-Herrera et al., 2017), these are; (i) discrete grain breakage (DGB) (Potapov and Campbell, 1994), (ii) the particle replacement model (PRM) (Cleary, 2001; Tavares et al., 2020) and (iii) the bonded particle model (BPM) (Potyondy and Cundall, 2004). The DGB and PRM models are semi-empirical, replacing individual discrete entity with a population of smaller entities in accordance with a prescribed appearance function. The BPM aims to directly simulate rock breakage with progeny sizes and shapes determined dynamically. Of the three methods, only the BPM accurately simulate crack propagation in accordance with fracture mechanics (Wang et al., 2000 and Potyondy and Cundall, 2004). These models are further reviewed in the following sub-sections

2.8.1   Discrete grain breakage model

Discrete grain breakage model (DGB) describes the interaction between polygons (2D) or polyhedral (3D) entities during stressing (Herbst and Potapov, 2004). A simplified version of DGB is the fast breakage model (FBM) which requires less computational effort and has the ability to handle a larger number of polyhedral entities (Potapov et al., 2007). A unique feature of this breakage model is that rock specimens are instantaneously broken into fragments through Laguerre-Voronoi tessellation. Breakage occurs when the total collision energy is greater than the required fracture energy. An illustration of this is shown in Fig. 2.10 where a rock specimen in a 2D simulation environment composed of polygons ($t = t_o$) is broken into fragments during stressing as the time progresses, from the left at $t = t_o + dt$ to the right at $t = t_o + 3dt$.

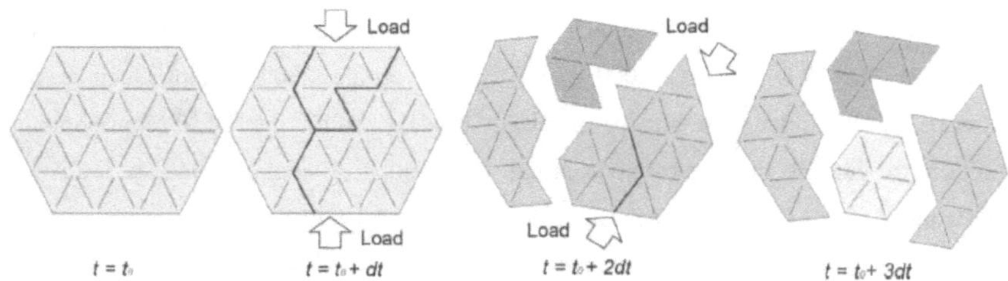

Fig. 2.10: An illustration of fast breakage model using a 2D rock specimen showing the fragmentation with time. Adapted from Jiménez-Herrera et al. (2017)

2.8.2 Particle replacement method

The particle replacement model (PRM) was pioneered by Cleary (2001). This model instantaneously replaces a discrete entity (any shape of rock) with a collection of smaller similar entities to form fragments once the defined failure criterion has been attained (Jiménez-Herrera et al., 2017; Tavares et al., 2020). An illustration of this model is shown in Fig. 2.11. Rock specimen exists as a discrete entity at time $t=to$ and begins to form a collection of discrete entities up to the application of the load at time $t = to+dt$. Fragmented products which do not experience any further load remain as discrete entities, while fragments that experience further load repeat the sequence of discrete entity replacement ($t = to + 2dt$).

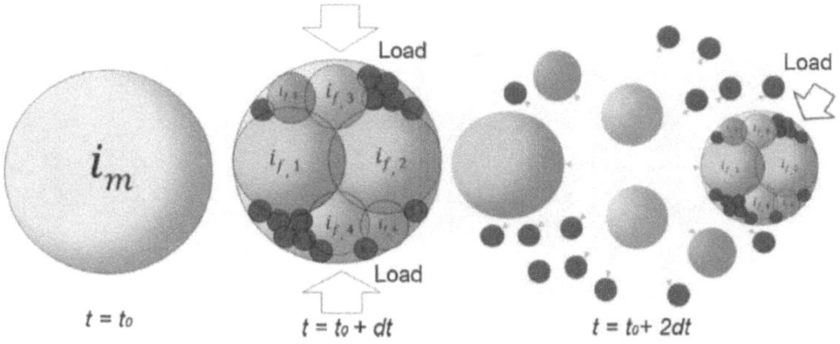

Fig. 2.11: An illustration of particle replacement method. Adapted from Jiménez-Herrera et al. (2017)

2.8.3 Bonded particle model

The Bonded particle model (BPM) represents individual rock specimen as an assembly of discrete elements, typically spheres denoted as DEM-spheres (Wang et al., 2000; Potyondy and Cundall, 2004; Potyondy, 2015). Adjacent DEM-spheres are initially connected via bonded (cohesive) interactions, similar to linear elastic springs. Through the action of gravity or moving walls, forces are applied to this bonded assembly causing mechanical stresses to accumulate between neighbouring spheres. A failure criterion governs the peak tensile (or compressive) stress within individual bonded interactions. Once the criterion is exceeded, the bonded interaction is removed and replaced with a cohesionless elastic repulsive interaction and frictional shear interaction. This approach simulates the formation of cracks within the rock specimen and permits subsequent frictional sliding along fracture surfaces so formed. Whilst the term "bonded particle model" was first coined by Potyondy and Cundall (2004), earlier examples of this approach exist in the literature, including the "lattice solid model" of Mora and Place (1994). These approaches of BPM by both authors have been found to reasonably simulate rock breakage. However, the mathematical approaches are slightly different from each other and are discussed in the following sub-sections.

2.8.3.1 Potyondy and Cundall's BPM approach

In this method, shown in a 2-D representation in Fig. 2.12, the interaction at the contact point between the connecting DEM-discs involves both normal interaction and the parallel bond interaction operating together to prescribe a change in force (Eqs 2.18 and 2.19) and change in moments to the connecting DEM-discs, shown in Eqs 2.20 and 2.21 (Potyondy and Cundall, 2004 and Potyondy, 2015). The contact interaction has both the normal and shear components of forces (Potyondy, 2015) while the parallel bond impact moments.

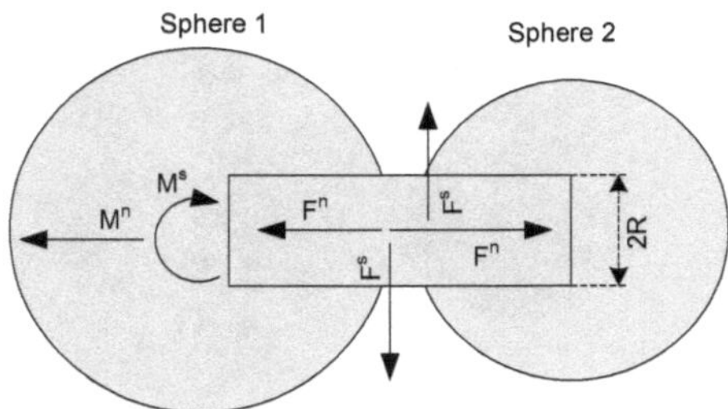

Fig. 2.12: (a) 2D -representation of Potyondy and Cundall BPM showing the forces and moments in the normal and shear direction (Jiménez-Herrera et al. 2017)

$$\Delta F^n = -k^n \, A \, v^n \, \Delta t \qquad\qquad 2.18$$

$$\Delta F^s = -k^s \, A \, v^s \, \Delta t \qquad\qquad 2.19$$

$$\Delta M^n = -k^s \, J \, \omega^n \, \Delta t \qquad\qquad 2.20$$

$$\Delta M^s = -k^n \, I \, \omega^s \, \Delta t \qquad\qquad 2.21$$

Where ΔF^n and ΔM^n are the force and moment respectively in the normal direction and ΔF^s and ΔM^s are the force and the moments respectively in the shear direction. k^n and k^s are the normal and shear stiffness of the contact bond respectively, v^n and v^s are normal and tangential velocities respectively ω^n and ω^s are normal and angular tangential velocities respectively. A, J and I are the cross-sectional area of the connecting beam, moments of inertia and polar moments of inertia are expressed in Eqs 2.22-2.24 respectively.

$$A = \pi R^2 \qquad\qquad\qquad 2.22$$

$$J = \frac{1}{2} \pi R^4 \qquad\qquad\qquad 2.23$$

$$I = \frac{1}{4} \pi R^4 \qquad\qquad\qquad 2.24$$

Where R is the radius of the connecting beam which is dependent on the diameter of the smallest DEM-disc.

Failure of a rock specimen occurs if either criterion given by Eqs 2.25 or 2.26 is satisfied, i.e., the computed normal stress (σ) exceeds the critical normal stress (σ_c) or shear stress (τ) exceeds the critical shear stress (τ_c) (Potyondy et al., 1996 and Donze et al., 1997).

$$\sigma = \frac{F_{Total}^n}{A} + \frac{2M^n}{J} R > \sigma_c \qquad\qquad\qquad 2.25$$

$$\tau = \frac{F_{Total}^S}{A} + \frac{2M^S}{I} R > \tau_c \qquad\qquad\qquad 2.26$$

2.8.3.2   Wang's BPM approach

This bonded particle model employs simple linear elastic springs between adjacent DEM-spheres, with failure occurring when the spring extension exceeds a prescribed distance (Mora and Place, 1994). This BPM simulates the failure of materials under pure tension but does not accurately predict crack propagation under compressive or shear loading. The latter is achieved by accounting for additional deformational degrees-of-freedom between adjacent DEM-spheres; specifically, shear, bending and torsional deformation in addition to tensile/compressive deformation.

Wang et al. (2006) notionally connect adjacent DEM-spheres with cylindrical linear elastic beams with six deformational degrees-of-freedom, as illustrated in Fig. 2.13. Each degree-of-freedom is mathematically represented via a Hookean interaction. Purely tensile (compressive) forces are proportional to the relative extension (compression) of adjacent DEM-sphere. Similarly, shear forces in each of the two directions perpendicular to the line joining the two sphere's centres-of-mass are calculated in proportion to the relative shear displacement. Ascribing an initial orientation to each DEM-sphere (rotational degrees-of-freedom in addition to translational) permits the calculation of relative bending and twisting between adjacent DEM-sphere; with bending and twisting moments calculated in proportion to the relevant changes in angle between adjacent DEM-spheres.

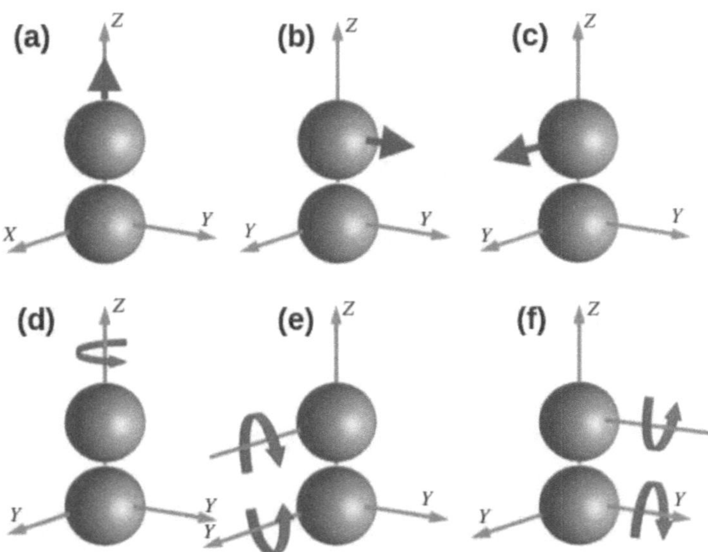

Fig. 2.13: The six interactions between bonded DEM-spheres: pulling or pushing in a radial direction (a), shearing forces in the tangential direction (b and c), twisting (d) and bending around two-axis (e and f) Wang et al., (2006).

With the assumption of isotropy of shearing and bending deformation, two forces (F_{ij}^{α}) (Eqs 2.27 and 2.28) and two moments (M_{ij}^{β}) (Eqs 2.29 and 2.30) are calculated between adjacent bonded pairs of DEM-spheres in any given time-step.

$$F_{ij}^{n} = K_{ij}^{n}\Delta U_{ij}^{n} \qquad\qquad 2.27$$

$$F_{ij}^{s} = K_{ij}^{s}\Delta U_{ij}^{s} \qquad\qquad 2.28$$

$$M_{ij}^{b} = K_{ij}^{b}\Delta\theta_{ij}^{b} \qquad\qquad 2.29$$

$$M_{ij}^{t} = K_{ij}^{t}\Delta\theta_{ij}^{t} \qquad\qquad 2.30$$

Where: ΔU_{ij}^{α} is the relative normal or shear ($\alpha = n\ or\ s$) displacement of adjacent spheres i and j, and $\Delta\theta_{ij}^{\beta}$ is the relative angle change due to bending or torsion ($\beta = b\ or\ t$) between the elements.

The four spring constants ($K_{ij}^{n}, K_{ij}^{s}, K_{ij}^{b}, K_{ij}^{t}$) are computed in accordance with linear elastic beam theory given by Eqs 2.31-2.34. Assuming adjacent spheres are connected via cylindrical beams whose cross-sectional radius is equal to the arithmetic mean of the radii of the elements [$R_{ij} = \frac{1}{2}(R_i + R_j)$] and whose equilibrium length is equal to the sum of the sphere radii, the following equations can be derived:

$$K_{ij}^n = \frac{\pi}{2}Y(R_i + R_j) \qquad\qquad 2.31$$

$$K_{ij}^s = \frac{\pi}{2}G(R_i + R_j) \qquad\qquad 2.32$$

$$K_{ij}^b = \frac{\pi}{8}Y(R_i + R_j)^3 \qquad\qquad 2.33$$

$$K_{ij}^t = \frac{\pi}{4}G(R_i + R_j)^3 \qquad\qquad 2.34$$

Where: Y is the interaction Young's modulus and $G = \dfrac{Y}{2(1+v)}$ is the interaction shear modulus; typically defined in terms of a dimensionless Poisson's ratio (v). It should be noted that these mechanical parameters are micro-physical quantities. The equivalent macroscopic mechanical properties are related to these micro-physical but, also depend on the topology of the assembly of bonded interactions comprising a DEM rock particle.

Failure of bonded interactions is governed by a Mohr-Coulomb failure criterion shown in Eq 2.35

$$\tau_{ij} \geq \begin{cases} C - \sigma_{ij}^n \tan\Phi & if\ \sigma_{ij}^n < C\tan\Phi \\ 0 & otherwise \end{cases} \qquad\qquad 2.35$$

Where: C is the cohesive strength of the interaction and Φ is the internal angle of friction. The normal stress (σ_{ij}^n) and shear stress (τ_{ij}) of the interaction are computed in accordance with linear elastic beam theory expressed as Eqs 2.36 and 2.37 respectively:

$$\sigma_{ij}^n = \frac{F_{ij}^n}{A_{ij}} + \frac{\left|M_{ij}^b\right|}{I_{ij}}R_{ij} \qquad\qquad 2.36$$

$$\tau^{ij} = \frac{\left|F_{ij}^s\right|}{A_{ij}} + \frac{\left|M_{ij}^t\right|}{J_{ij}}R_{ij} \qquad\qquad 2.37$$

Where: $A_{ij} = \pi R_{ij}^2$ is the cross-sectional area, $I_{ij} = \pi R_{ij}^4/4$ is the bending moment of inertia and $J_{ij} = \pi R_{ij}^4/2$ is the polar moment of inertia of the interaction connecting DEM-spheres i and j.

2.9 Chapter summary

Rock breakage finds application in many fields including the comminution stage in mineral processing. The energy associated with the size reduction of rocks during comminution is commonly determined to be relatively high. Many studies have thus been devoted to understanding breakage mechanisms in order to work toward improving the process.

Breakage mechanisms can be traced to back to the fundamentals of rock and fracture mechanics. In rock mechanics, the standard uniaxial compressive breakage test shows that rock and rock-like materials exhibit complex behaviour which can be described at different stages of loading. These are; crack closure, elastic deformation, elastic limit, stable and unstable crack extension ending in failure. Also, in fracture mechanics, studies of deformation and failure of load-bearing materials such as rocks on a microscopic scale have developed from Griffith's theory. This examines the concepts of crack initiation and propagation as well as the concept of energy balance. Studies mainly focus on the initiation and extension of cracks in length in a manner almost parallel to the direction of the applied load.

There are various techniques of rock specimen characterisation that span across the field of rock and fracture mechanics as well as mineral processing. However, among these techniques, the single rock specimen breakage test has been regarded as the most basic technique through which fundamental understanding of breakage can be developed.

A majority of comminution research has been conducted using single rock specimen breakage tests. This has facilitated the development of various laboratory-scale comminution testing devices used to quantify the force and energy requirements to achieve a target product size as well as to determine material characterisation properties. The impact load cell is one of such devices which combines advantages of both the drop weight tester and split-Hopkinson pressure bar. The force-time history obtained from the impact event in this device allows for the inference of several fracture measures and material properties, along with a qualitative and quantitative review of the extent of breakage.

Several factors have been identified to affect breakage behaviour of rock specimens during breakage tests Attributing discrepancies in breakage test results to a definite single factor remains a challenge. Current approaches to breakage testing have mostly been making use of real rock specimens which have inherent flaws (pre-existing cracks) and often little-known mineralogical composition, making results difficult to replicate. However, an alternative approach to breakage testing using synthetic rocks has emerged which provides controlled material characteristics akin to real rocks, with minimal flaws and less sample heterogeneity. While the use of synthetic rock specimens on the laboratory-scale breakage device is valuable, it is nevertheless impractical to determine micro-scale detail such as the rock specimen's internal stress propagation and crack propagation. Numerical methods such as the discrete element method (DEM) have the capability to extend knowledge in this area.

While DEM was initially designed to study granular flow, its application has since been extended to study dynamic scenarios such as breakage under high strain rates, including areas such as earthquake studies, seismic failure and ore breakage in comminution devices. To facilitate this, different breakage models have been developed for use in DEM, such as the discrete grain breakage (DGB) the particle replacement model (PRM) and the bonded particle model (BPM). Among those reviewed, the bonded particle model (BPM) has been shown to be the most suitable to simulate crack propagation in accordance with fracture mechanics. Distinct forms of BPM have been identified in literature which varies in their mathematical formulations. The approach developed by Wang et al. (2006) employs simple linear elastic springs between adjacent elements, with failure occurring when the spring extension exceeds a prescribed distance. This form has been selected for this work.

In conclusion, this study will focus on using a BPM-DEM approach to investigate in isolation, the load response on rock specimens with varying levels of pre-existing cracks and mineralogical composition. To achieve these, a numerical methodology of a single rock specimen breakage by impact mechanism was developed. The BPM in DEM was employed to simulate the fracture mechanics of rock specimen breakage and their relation to the typically measured fracture force and pattern from a SILC experiment. The following chapter summarizes the methodology that was followed.

3 Calibration of bonded particle model mechanical properties

3.1 Overview

In numerical simulations, the mechanical behaviour of a rock specimen is governed by the specification of the microstructural model parameters against measurable macroscopic mechanical properties (Yoon, 2007; Potyondy, 2015; Han et al., 2017). The number of specifications, as well as the mathematical relationships between model parameters and the measured macroscopic parameters of BPMs, has been found to vary slightly from one DEM package to another (Fakhimi, 1998; Itasaca, 1999; Wang and Tonon, 2010; DEM-solutions, 2011). However, there are often similarities in their implementation or simulation results. The International Society of Rock Mechanics (ISRM) and American Society for Testing and Materials (ASTM) geotechnical standard technique (International Society of Rock Mechanics, 1979; 2007; American Society for Testing and Materials, 1991; 2013) are global geotechnical standards that are usually adopted to calibrate relevant model parameters against four main macro mechanical properties of a rock specimen. These macroscopic mechanical properties are; Young's modulus (Y), Poisson's ratio (v), Unconfined compressive strength (UCS) and Unconfined tensile strength (UTS).

Four main approaches have been identified for most BPM calibrations against the measured macroscopic mechanical properties specified:

- The first and most common approach is by trial and error where model parameters are tuned/varied to match an actual macroscopic parameter. However, this becomes more complicated as the number of model parameters increase (Zhang, 2010).

- The second approach is the Design of Experiment (DoE) which is usually employed when there are many model parameters to the calibrated. This often firstly involves screening out the statistically insignificant parameters and thereafter, calibrating and optimising the significant parameters against the measured macroscopic parameters (Johansson et al., 2016; Chehreghani et al., 2017).

- The third is the machine learning or neural network approach. Here, the unknown macroscopic property is predicted using statistical interpolation or a search pattern algorithm to identify model parameters from a database of properties and thereafter matched to a macroscopic property (Wang and Cao, 2017; Alnaggar and Bhanot, 2018; Zhai et al., 2019)

- The last approach is known as the systematic parameter sensitivity study which is employed in the current work. This approach is similar to the DoE approach but entails systematically varying the model parameters against macroscopic parameters. Particular measured parameters are subsequently fitted to a function to describe a relationship between the model parameter and macroscopic properties.

The numerical procedure adopted for the measurement of these macroscopic parameters according to the set standard and the application of the systematic parametric approach to the current study are described in subsequent sections.

Numerical simulations were conducted using ESyS-particle 2.3.5, an open-source DEM package which uses Python-based libraries to generate geometries and simulations. A volume filling algorithm was used to populate the DEM elements for the prescribed shape (Abe and Mair, 2005). Adjacent spheres were recognised via a neighbour search algorithm and connected via brittle-elastic bonds (Place and Mora, 2001). The forces and moments experienced upon deformation were determined according to the description in section 2.8.3.2. Simulations were performed using High performance computers provided by the University of Cape Town's ICTS Computing team (http://hpc.uct.ac.za) which allows parallel simulations of jobs. Details of the computational resources that were allocated are as follows: cluster type Dell C6420, 40 cores, 384GB RAM and a computation time limit of 170 hours.

3.2 DEM measurement of macroscopic mechanical properties

A suite of uniaxial compression and direct tension simulations were conducted to demonstrate and measure the macroscopic mechanical properties of the bonded particle DEM model. Simulations involved the construction of a cylindrical specimen of bonded DEM-spheres with an overall diameter and length of 10mm and 20mm, respectively. The range of DEM element radii was kept to the same as that of the SILC models in this work: between 0.08mm and 0.32mm. A typical DEM specimen assembly for these simulations is shown in Fig. 3.1.

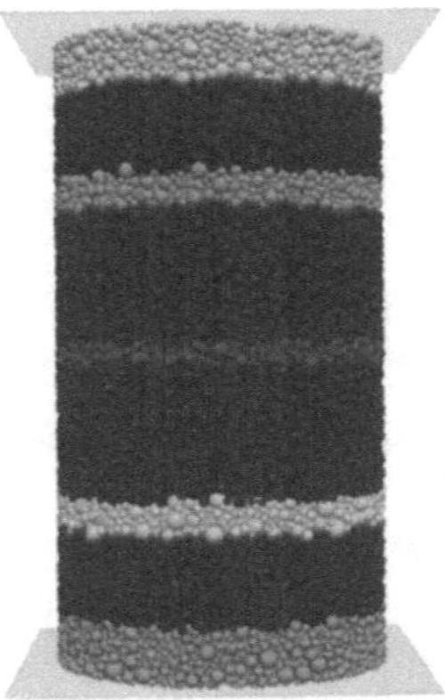

Fig. 3.1: Typical DEM specimen assembly used for uniaxial compression and direct tension simulations.

DEM-spheres within 1.5mm of each end of the cylinder were connected to planar walls via linear elastic springs that imparted forces only in the vertical loading direction. During simulations, these walls were moved towards (or away from) one-another, to impart compressive (or tensile) stress to the cylinder. After an initial period of 10000 time-steps, during which the speed of the walls accelerated at a constant rate, the walls moved at a fixed relative velocity of 0.25m/s. The net force acting between the walls and spheres bound to them was continuously recorded throughout the simulations. The instantaneous axial stress was then obtained by dividing these forces by the initial cross-sectional area of the cylinder.

To determine the macroscopic Young's modulus of the specimen, it was also necessary to monitor the axial strain within the specimen. Thus, the mean vertical position was continuously monitored for two layers of DEM-spheres at 25% and 75% of the cylinder height, illustrated in Fig. 3.1. The relative vertical separation between these layers divided by their initial separation determined the axial strain at any given instant during the simulations.

Finally, to calculate the macroscopic Poisson's ratio of the specimen, a measure of the diametrical strain was required. Coordinates of a ring of surface spheres surrounding the mid-height of the cylinder were recorded. The mean diameter of this ring was then determined at every time interval to obtain a measure of the instantaneous diametrical strain.

Fig. 3.2 shows typical axial stress vs. axial strain and axial stress vs. diametrical strain plot obtained for uniaxial compression/tension simulation. In accordance with standard geotechnical testing practise in ISRM standards (Bieniawski and Hawkes, 1978), the slope of the linear section of the axial stress-axial strain curve is the macroscopic Young's modulus. The peak axial stress is the unconfined compressive strength (UCS (σ_c)) of the specimen while the peak axial stress under tensile loading gives the unconfined tensile strength (UTS (σ_t)). The slope of the linear section of the axial stress-diametrical strain curve determines the macroscopic Poisson's ratio when divided by Young's modulus.

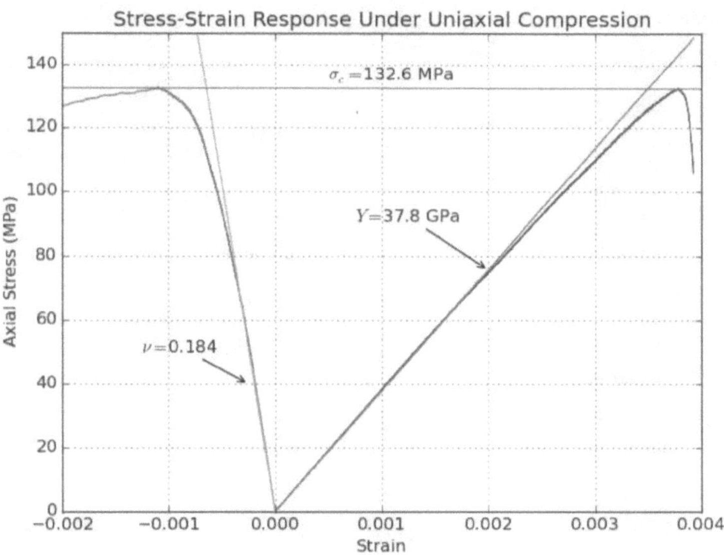

Fig. 3.2: Axial stress vs. axial strain (blue plot) and axial stress vs. diametrical strain (green plot) curves for a typical uniaxial compression simulation with an illustration of the calculated Young's modulus, Poisson's ratio and compressive strength

3.3 Development of empirical calibration relationships

For the BPM formulation employed herein, the macroscopic mechanical properties of a bonded DEM-spheres assembly were calibrated via prescription of the values of four mechanical model parameters; elastic beam modulus (Y_b), beam Poisson's ratio (ν_b), beam cohesive strength (C_b) and tangent of the beam internal angle of friction ($\tan \theta$). The first two parameters govern the elastic response of the DEM-sphere assembly, while the latter two govern the tensile and compressive strength of the assembly. These four parameters may be calibrated to produce DEM rock specimens with macroscopic mechanical properties matching those measured for natural rock specimens via geotechnical characterisation tests (Bieniawski and Hawkes, 1978; International Society for Rock Mechanics, 2007).

In this systematic parameter calibration study approach, a set of default parameters (Table 3.1) were initially chosen following which parameters were varied within a specified range.

Table 3.1: Default model parameters

Model Parameter	Value
Y_b	75 GPa
v_b	0.25
C_b	175.0 MPa
$\tan\theta_b$	1.0
Friction coefficient	0.6
viscosity	$0.0 \ \mathrm{kg.m^{-1}s^{-1}}$
Sample height	20.0 mm
Sample radius	5.0 mm
Maximum sphere radius	0.32 mm
Minimum sphere radius	0.08 mm
Plate speed	$0.25 \ \mathrm{ms^{-1}}$

Model parameters were paired and systematically varied against a measured macroscopic property. The beam elastic modulus and beam Poisson's ratio were paired and varied against the measured macroscopic Young's modulus obtained under compressive and tensile loading. The cohesive strength and tangent of the internal angle of friction were paired and varied against the measured UCS and UTS. Table 3.2 summarises the ranges used to vary the model parameters, the interval of variation for a successive increment, as well as the total number of levels of the individual model parameters.

Table 3.2: Range of values of the model parameter

Model Parameter	Range of values	Interval	Number of levels
Y_b	25 – 125 (GPa)	25 GPa	5
v_b	0.1 – 0.9	0.2	5
C_b	25 – 325 (MPa)	50 MPa	7
tanθ	0.1 – 2.1	0.5	5

Figs 3.3 and 3.4 show the correlation between the macroscopic Young's modulus under compression/tension and beam elastic modulus for each value of Poisson's ratio. In both graphs, it can be observed that a linear relationship exists between Young's modulus and elastic beam modulus but with different slopes at each value of Poisson's ratio. This implies that the measured Young's modulus depends on both the elastic beam modulus and the beam Poisson's ratio.

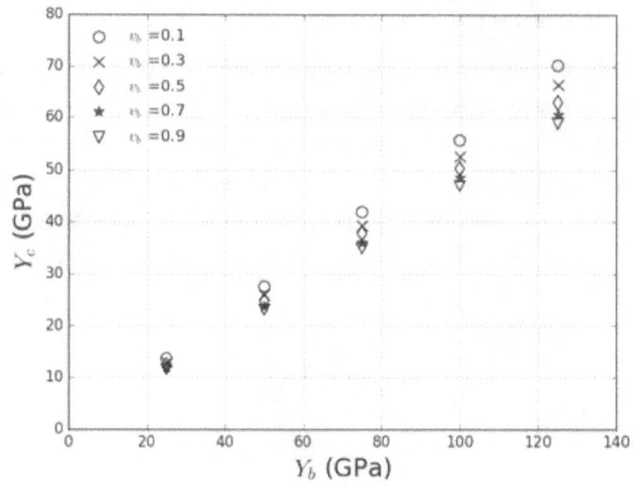

Fig. 3.3: Macroscopic Young's modulus under compressive loading (Y_c) versus elastic beam modulus (Y_b) of bonded DEM-spheres, for a range of beam Poisson's ratios

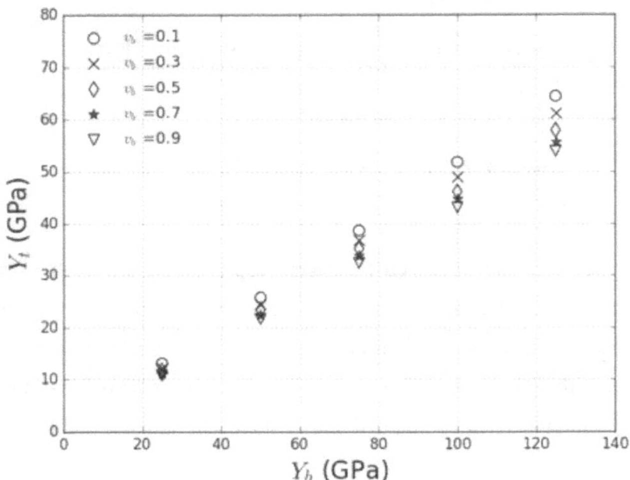

Fig. 3.4: Macroscopic Young's modulus under compressive loading (Y_t) versus elastic beam modulus (Y_b) of bonded DEM-spheres, for a range of beam Poisson's ratios.

In order to eliminate the linear dependency of the elastic beam modulus and study the dependency of beam Poisson's ratio in isolation, the measured Young's modulus under compression and tension were divided by the beam elastic modulus. These ratios were then plotted against the beam Poisson ratio, which gives collapse data points, as shown in Figs 3.5 and 3.6. It can be observed that the Poisson's ratio showed an inverse and "monotonic-linear" relationship with these ratios, i.e., the relationship is non-linear or linear with a slight curvature.

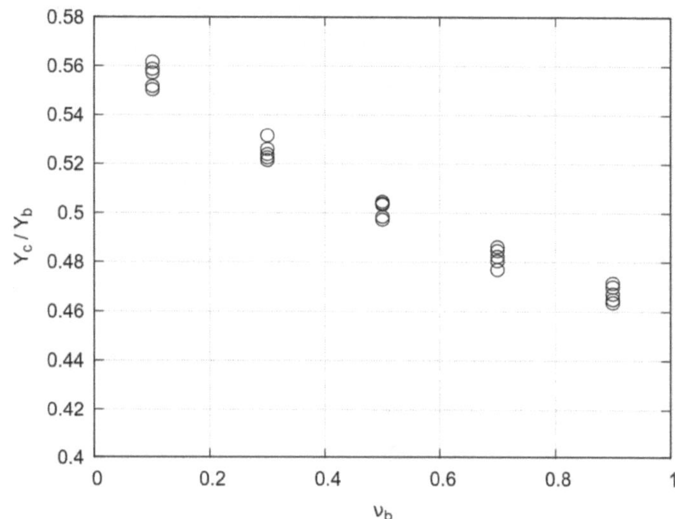

Fig. 3.5: the ratio of Young's modulus obtained under compression to the elastic beam modulus (Y_c/Y_b) versus the beam Poisson's ratio (v_b)

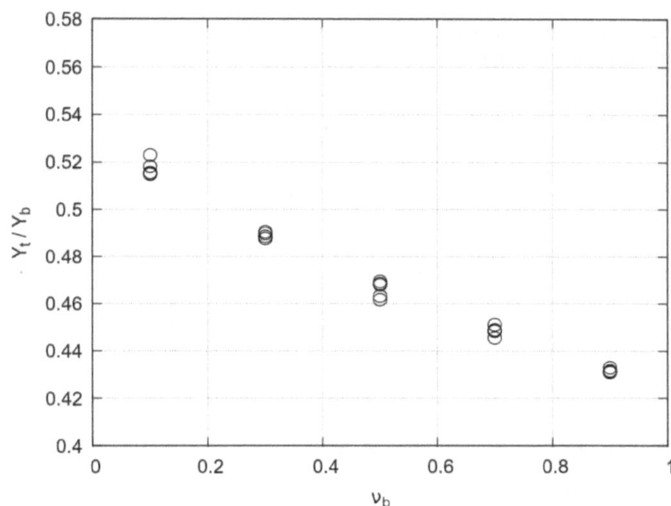

Fig. 3.6: The ratio of Young's modulus obtained under tension to the elastic beam modulus (Y_t/Y_b) versus the beam Poisson's ratio (v_b)

Similarly, the dependencies of the UCS and UTS on the cohesive strength at different values of the internal angle of friction are plotted in Figs. 3.7 and 3.8. A linear relationship can be observed between the UCS or UTS and cohesion at different internal angle of frictions.

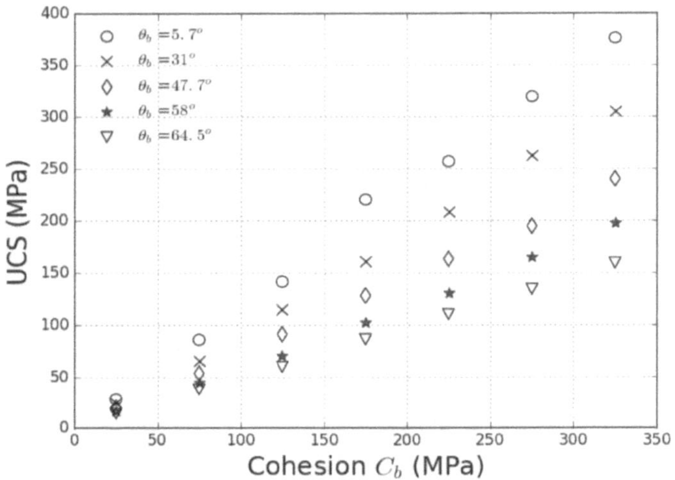

Fig. 3.7: Macroscopic UCS versus microscopic cohesive strength (C$_b$) of bonded DEM-spheres, for a range of beam bond internal angle of frictions.

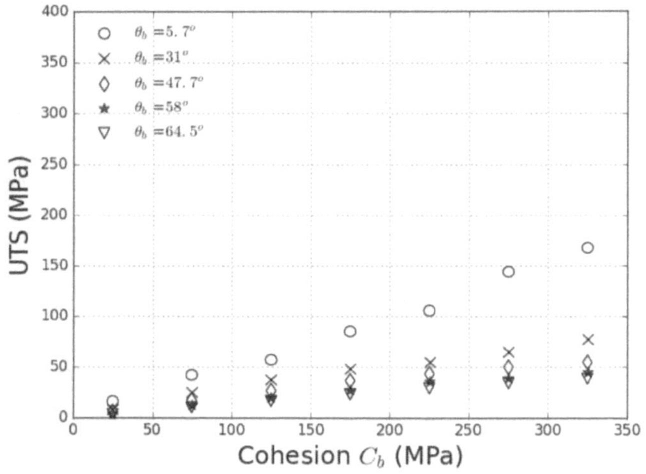

Fig. 3.8: Macroscopic UCS versus microscopic cohesive strength (C$_b$) of bonded DEM-spheres, for a range of beam bond internal angle of frictions.

Further collapsing the data points by dividing the UCS and UTS by cohesion and plotting these ratios against the internal angle of friction, the variation of the friction angle on the UCS and UTS becomes more apparent with an inverse monotonic trend as shown in Figs 3.9 and 3.10 respectively (Han et al., 2017).

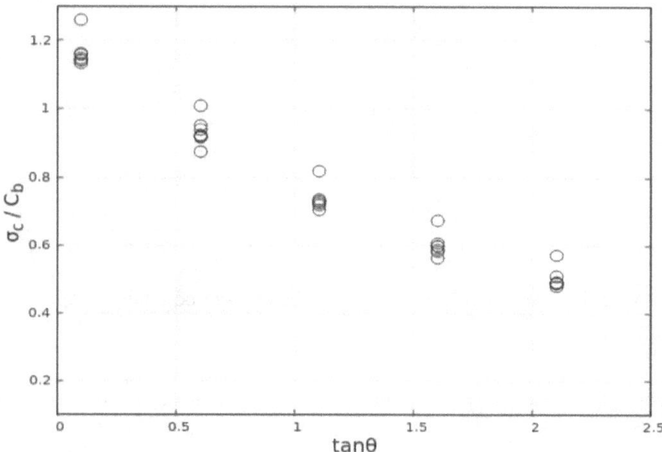

Fig. 3.9: The ratio of unconfined compressive strength to the beam cohesive strength (σ_c/C_b) versus the tangent of the internal angle of friction (tanθ)

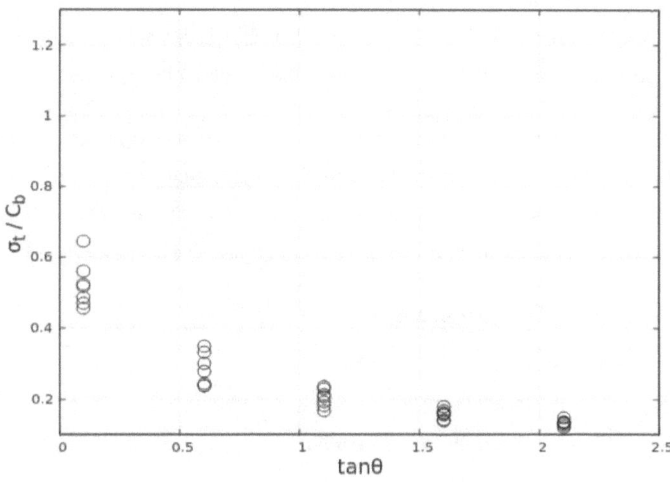

Fig. 3.10: The ratio of unconfined tensile strength to the beam cohesive strength (σ_t/C_b) versus the tangent of the internal angle of friction (tanθ)

Having examined the trend and form of relationship between the model and macroscopic parameters, empirical relationships were proposed based on similar prior calibration studies (Han et al., 2017; Weatherley, 2013). Fitting functions as expressed in Eqs. 3.1-3.4 were thus used to relate the macroscopic properties to paired model parameters.

$$Y_c = \frac{K_1 Y_b}{(1+K_2 v_b)} \qquad\qquad 3.1$$

$$Y_t = \frac{K_3 Y_b}{(1+K_4 v_b)} \qquad\qquad 3.2$$

$$UCS\ (\sigma_c) = \frac{K_5 C_b}{K_6 + \tan\theta} \qquad\qquad 3.3$$

$$UTS\ (\sigma_t) = \frac{K_7 C_b}{K_8 + \tan\theta} \qquad\qquad 3.4$$

Where: Y_c and Y_t (GPa) are the macroscopic Young's modulus measured under compression and tension respectively.

v_c and v_t are the macroscopic Poisson's ratio measured under compression and compression respectively.

UCS or σ_c and UTS or σ_t (MPa) are the unconfined compressive strength and unconfined tensile strength respectively.

K_i (i = 1, 2, …8) are constants determined by the empirical fit which depend on the spatial relationship of the connecting adjacent DEM-spheres within the bonded rock specimen.

Non-linear regression of the simulation data points against the model was conducted using the SciPy optimise curve fit Python package which returned the calculated fit parameters (K-values) and the estimated covariance as summarised in Table 3.3. The goodness of fit indicated by the high R-squared values suggest that these correlations are acceptable.

Table 3.3: Summary of K values obtained from models fit

K-value	K-value	R^2
$K_1 = 0.568 \pm 0.00187$	$K_2 = 0.236 \pm 0.00612$	0.9996
$K_3 = 0.526 \pm 0.00135$	$K_4 = 0.245 \pm 0.00480$	0.9998
$K_5 = 1.850 \pm 0.0470$	$K_6 = 1.46 \pm 0.0491$	0.9956
$K_7 = 0.286 \pm 0.0110$	$K_8 = 0.471 \pm 0.0260$	0.9868

These constants were substituted into Eq. 3.1 to 3.4 establish the relationships between the model and macroscopic parameters proposed as a basis for calibration of rock specimens. Fig. 3.11 (a-d) show the summary of these empirical relationships fitted into collapsed data points.

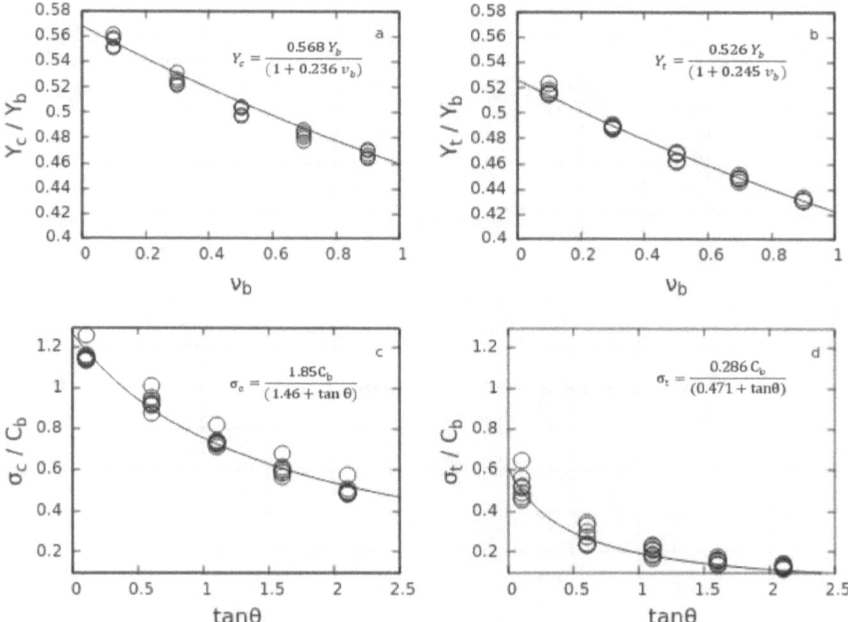

Fig. 3.11: (a) the ratio of Young's modulus obtained under compression to the elastic beam modulus (Y_c/Y_b) versus the beam Poisson's ratio (v_b) with fitting curve plotted with Eq. 3.1. (b) The ratio of Young's modulus obtained under tension to the elastic beam modulus (Y_t/Y_b) versus the beam Poisson's ratio (v_b) with fitting curve plotted with Eq. 3.2. (c) The ratio of unconfined compressive strength to the beam cohesive strength (σ_c/C_b) versus the tangent of the internal angle of friction ($\tan\theta$) with fitting function plotted with Eq. 3.3. (d) The ratio of unconfined tensile strength to the beam cohesive strength (σ_t/C_b) versus the tangent of the internal angle of friction ($\tan\theta$) with fitting function plotted with Eq. 3.4.

Finally, Figs. 3.12 to 3.15 show the overall summary of the simulation data points depicted as surface plots as well as their respective model functions fitted into the surface plot. These are to illustrate the applicable range of these DEM calibration relationships.

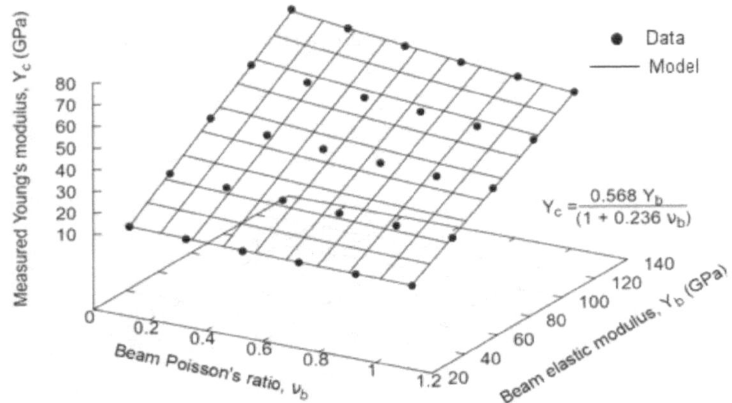

$$Y_c = \frac{0.568\ Y_b}{(1 + 0.236\ v_b)}$$

Fig. 3.12: Surface dot-plot and model fit of calibration relationship between measured Young's modulus under compression at different beam elastic modulus and beam Poisson's ratio

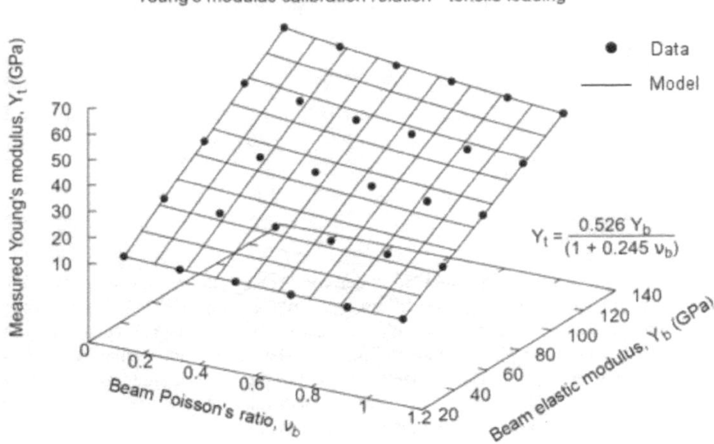

$$Y_t = \frac{0.526\ Y_b}{(1 + 0.245\ v_b)}$$

Fig. 3.13: Surface dot-plot and model fit of calibration relationship between measured Young's modulus under tension at different beam elastic modulus and beam Poisson's ratio

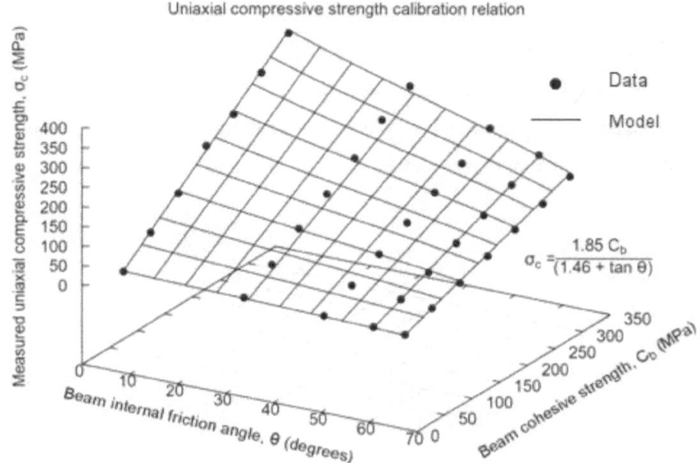

$$\sigma_c = \frac{1.85\, C_b}{(1.46 + \tan\theta)}$$

Fig. 3.14: Surface dot-plot and model fit of calibration relationship between measured UCS at different beam cohesive strengths and internal angle of frictions.

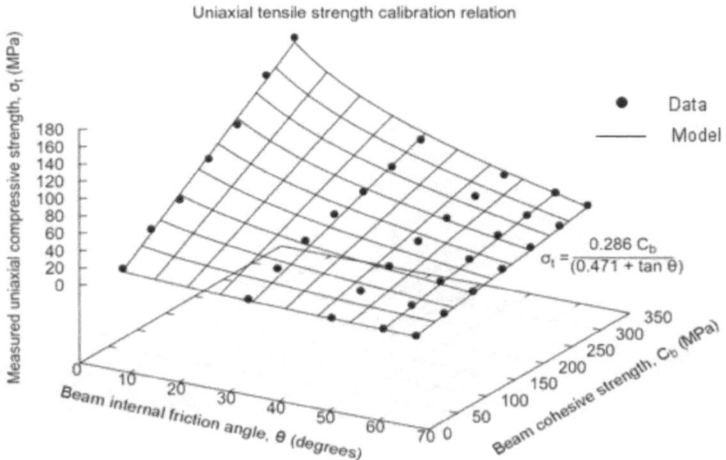

$$\sigma_t = \frac{0.286\, C_b}{(0.471 + \tan\theta)}$$

Fig. 3.15: Surface dot-plot and model fit of calibration relationship between measured UTS at different beam cohesive strengths and internal angle of frictions.

3.4 Chapter summary

An overview of different techniques to calibrate model parameters to obtain material properties according to ISRM and ASTM has been presented. The systematic parameter sensitivity study employed in this current study was used to calibrate and develop empirical calibration relationships to relate model parameters to macroscopic material properties.

4   DEM SILC model methodology and analysis

This chapter is an excerpt from an article (Oladele et al., 2019) that has been submitted for publication and currently under review.

4.1   Numerical SILC setup

The volume filling algorithm (see Section 3.1) was similarly used to populate the DEM elements for the prescribed shape in the SILC simulations. A horizontal wall with an upward active plane was specified to represent the top of the steel rod in a SILC device, and a sphere configured with the properties of steel was specified as the impacting steel ball. The force on the rock specimen was then calculated at discrete time intervals from the resultant force between the steel ball and bottom wall. Figs 4.1a and b show a snapshot of the simulation as well as an illustration of the force-time histories measured for 10 different samples.

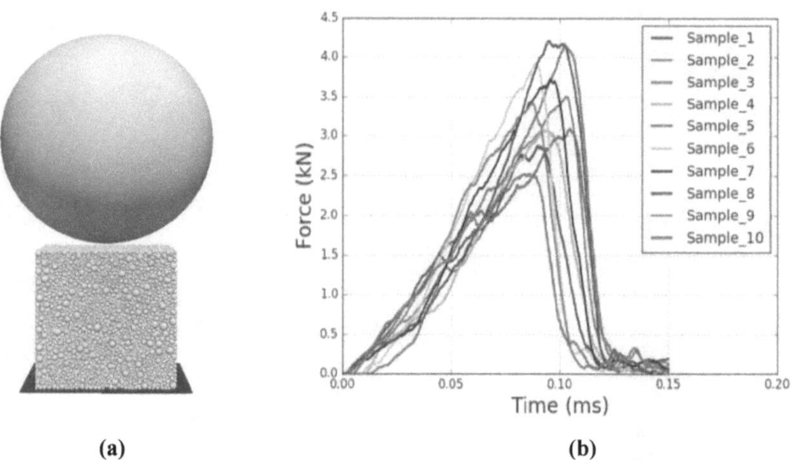

(a)   (b)

Fig. 4.1: Numerical SILC setup (a) and force-time histories for 10 different samples (b)

4.2   Measurement of fracture characteristics

Similar to SILC experiments, data measured from the numerical SILC simulations were used to calculate and quantify fracture characteristics of rock specimens.

4.2.1   Fracture force

The most important fracture characteristic usually obtained from the impact breakage of rock specimen in a SILC experiment is the fracture force. This is the peak (F_{peak}) measured from the force-time or force-displacement history of the impact event (Bourgeois and Banini, 2002; Tavares and King, 2004 and Bonfils, 2017). This definition is also used to quantify the fracture force in this study. The peak or fracture force for each simulation was calculated by first passing the instantaneous force profile for the steel bar through a low pass filter to remove high-frequency numerical noise. Thereafter the peak value of the filtered force time-history was calculated, thus ensuring that the variability of peak force estimates was not encumbered by numerical noise. An example of both an unfiltered and filtered force time-history is shown in Fig. 4.2.

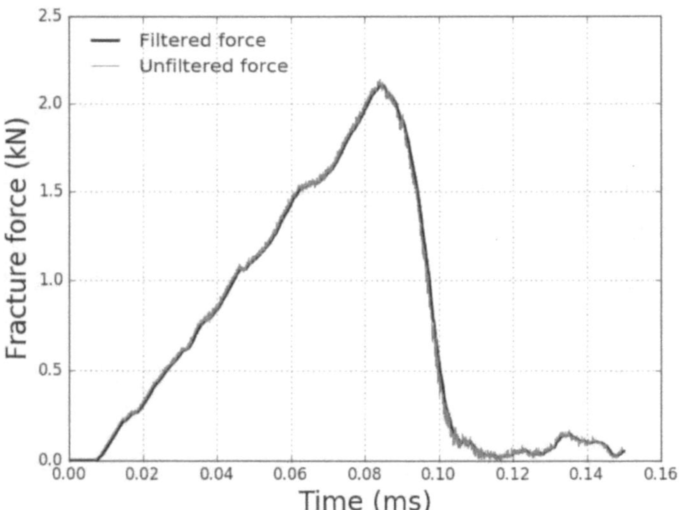

Fig. 4.2: Graph showing an illustration of the application of a low pass filter to remove numerical noise from simulated force time-histories

4.2.2   Linear elastic stiffness

The stiffness of a material is its resistance to deformation, which is usually expressed as the ratio of the applied force (F) to the deformation (d) produced by the force. Similarly, for the rock specimen in this study, the stiffness is the slope in the elastic regime of the force-displacement curve. This is illustrated in Fig. 4.3. with a straight line fit over the elastic regime, with the slope calculated to give the rock specimen stiffness (k), by Eq. 4.1. In terms of the macro-mechanical property i.e., Young's modulus, it is given by Eq. 4.2

$$k = \frac{\Delta F}{\Delta d} \qquad\qquad 4.1$$

$$k = \frac{YA}{L} \qquad\qquad 4.2$$

Where Y is the elastic or Young's modulus, A is the cross-sectional area and L is the length of the specimen.

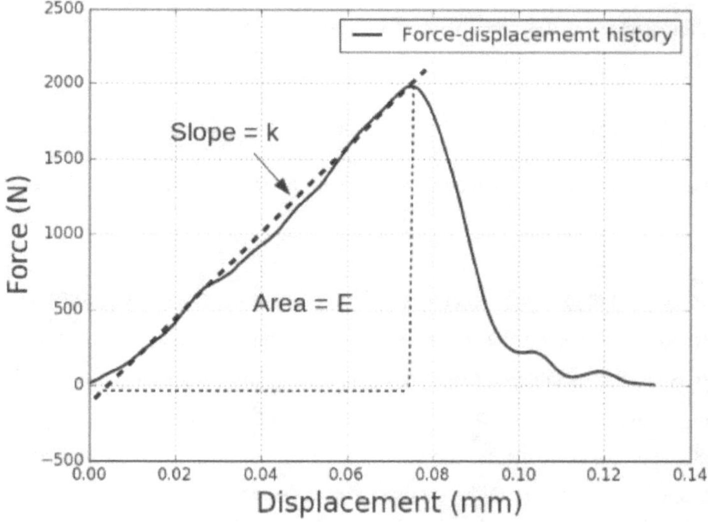

Fig. 4.3: Illustration of the methodology used to determine elastic stiffness from a force-displacement curve.

4.2.3   Fracture energy and fracture energy distribution

The fracture energy of a rock specimen is the energy that is stored in the specimen up to the point of fracture (Baumgardt, 1975 and Tavares, 2007) and is usually estimated to be the area under the force vs. deformation curve (Eq 4.3).

$$E = \int_0^{\Delta_f} Fd\Delta \qquad\qquad 4.3$$

Where Δ and Δ_f are rock deformation and deformation at fracture respectively. Assuming the linear-elastic portion of the force-deformation graph to be linear, the fracture energy can be simplified to the area of a triangle, as given by Eq. 4.4. In this scenario, the material is assumed to behave with linear stiffness up to the point of fracture. The specific fracture energy was obtained by dividing the fracture energy by the specimen's mass.

$$E = 0.5 * \Delta_f * F_{peak} \tag{4.4}$$

Variability of the fracture energy for different rock specimens can be estimated using probability distributions, such as the normal, log-normal or Weibull distribution (Baumgardt *et al.*, 1975; King and Bourgeois, 1993; Tavares, 2007 and Bonfils, 2007). For this work, the variability of fracture energy is described using the log-normal distribution function (*P(E)*), given by Eq. 4.5.

$$P(E) = \frac{1}{2}[1 + erf(\frac{ln\,E - ln\,E_{50}}{\sqrt{2}\sigma})] \tag{4.5}$$

Where E_{50} and σ are the median and standard deviations of the fracture energy respectively.

4.2.4 Stress tensor

The distribution of stress within a rock specimen under load does not always lead to fracture. Under an applied load, it is known that stress is induced within the specimen, resulting in fracture when it exceeds the specimen strength. Examination of this precursor during loading is significant. Among the methods of representing this distribution and/or the magnitude of stress is the stress tensor. To compute stress tensor, each specimen was initially divided into voxels containing approximately 100 DEM-spheres. The stress tensor was then computed as the average stress of each group of DEM-spheres in each voxel, given by Eq. 4.6 (Potyondy and Cundall, 2004).

$$\bar{\sigma}_{ij} = -\left(\frac{1-\phi}{\sum_{n_p} V^{(s)}}\right) \sum_{n_p} \sum_{n_c} \left|x_i^{(c)} - x_j^{(s)}\right| n_i^{(c,s)} F_j^c \tag{4.6}$$

Where $\bar{\sigma}_{ij}$ is the average stress in a voxel, ϕ is the porosity within the voxel of measurement, n_p and n_c are the number of DEM spheres and contacts within the voxel of measurement respectively, $V^{(s)}$ is the volume of DEM sphere, $x_j^{(s)}$ and $x_i^{(c)}$ are the location of a centre of a DEM sphere and its contact respectively, $n_i^{(c,s)}$ is the unit normal vector from the centre of a DEM sphere directed to its contact location and F^c is the force acting at contact.

The stress tensor on the rock specimen was visualised in a post-process analysis within the simulation domain.

4.3 Simulated breakage behaviour during a single impact event

A case study was considered where simulations were conducted on a representative 10mm cylindrical rock specimen. The macro-scale responses were extracted and studied in relation to their microscale behaviour. Among those investigated are the force-time history, new free surfaces generated based on the percentage of broken bonds (beams joining DEM spheres together), stress (stress tensor within the simulation domain) and energy distribution within the rock specimen and fragmentation. The relevant parameters used for simulations are summarised in Table 4.1. These are similar parameters used for the comparative study in chapter 6. Fig. 4.4 provides a comparison plot of the force, percentage of broken bonds and the energy across the duration of the impact. The behaviour of the force history showed similar behaviour to those typically observed in SILC experiments (discussed in section 2.3.3). Other key behaviours with respect to the breakage of bonds are discussed at four different time zones (A, B, C and D).

Table 4.1: Parameters used for the DEM-SILC breakage study

Parameter	Notation	Value
Model resolution	R_{max}/L	0.032
Beam elastic modulus	Y_b	1.8 GPa
Cohesion	C	37.23 MPa
Beam Poisson's ratio	v_b	0.25
Tangent of the beam internal angle of friction	$\tan\theta$	1
Steel ball radius	R_s	25.4mm
Drop height	H	100mm

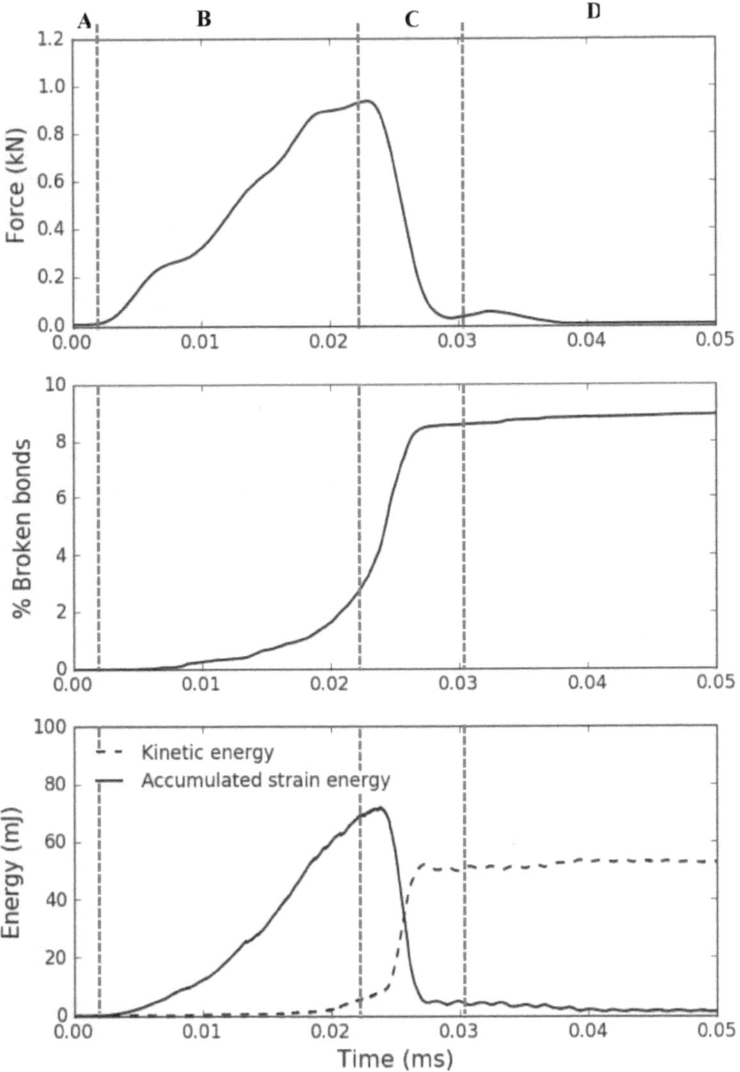

Fig. 4.4: A plot of the force, percentage of broken bond and energy against time. Time zones A (prior to impact), B (Prior to mechanical failure), C and D apply to all three graphs

4.3.1 Prior to impact

The prior to impact stage covers time zone A which is from 0 to around 0.002ms. This period is when the steel ball has yet to collide with the rock specimen, with all the bonds between DEM-spheres intact. During this period, no force is recorded, nor are there any broken bonds or energy input to the rock specimen from the impactor. This manifests as negligible force values recorded over the interval in Fig. 4.4.

4.3.2 Prior to mechanical failure

Referring to Fig. 4.4, this period covers time zone B and ends before the peak point of the force-time history at approximately 0.023ms. Three points are highlighted in discussing the events of this period; 0.003, 0.01 and 0.02ms which roughly correspond to the start, mid and endpoints respectively of this time zone. The steel ball initiates impact with the rock specimen at about 0.003ms. The measured force consequently shows a gradual increase with time as stress intensifies within the specimen. It is notable that the force slightly oscillates as it progresses upward. This is typical of SILC experiments and attributed to the propagation of elastic strain waves transferred through the rock specimen to the steel rod. As the stress intensifies, there is a progressive accumulation of damage shown by the ramping increase in the percentage of broken bonds.

The accumulated strain energy, or elastic energy stored within the rock specimen, also increases over this period. Stress tensor diagrams at the indicated times are given in Figs 4.5a, 4.6a and 4.7a. These demonstrate the manner in which the interior of the rock specimen is subjected to increasing stress which causes strain and bond breakage. This consequently results in the generation of fragments, whose formation can be found by identifying locations where there is high bond breakage. Figs 4.5b, 4.6b and 4.7b show such fracture locations initiating from the impact zone and the formation of a fragment which severs from the main rock specimen at 0.02ms as shown in Fig. 4.7c. Time zone B is therefore categorised as a damage zone.

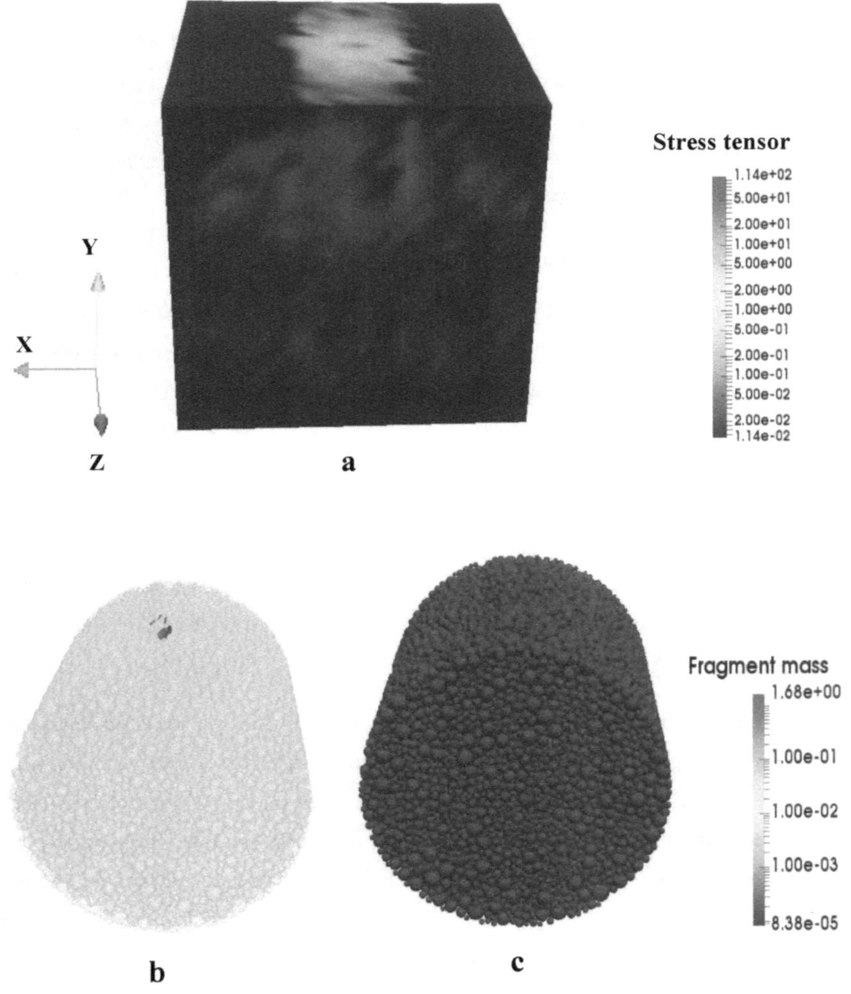

Fig. 4.5: Load response at 0.003ms. (a) Stress distribution   (b) Fracture locations within the rock specimen, with DEM-spheres, made slightly invisible for clear visualisation (c) Fragment formation

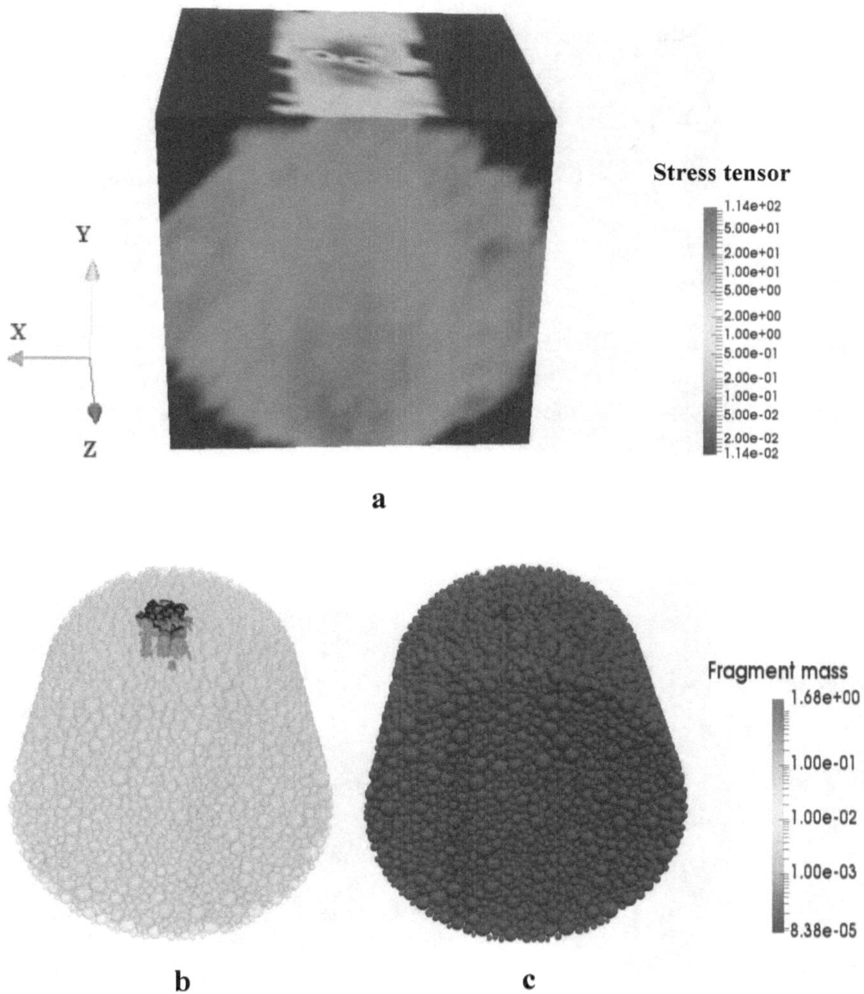

Fig. 4.6: Load response at 0.01ms. (a) Stress distribution   (b) Fracture locations within the rock specimen, with DEM-spheres, made slightly invisible for clear visualisation. (c) Fragment formation

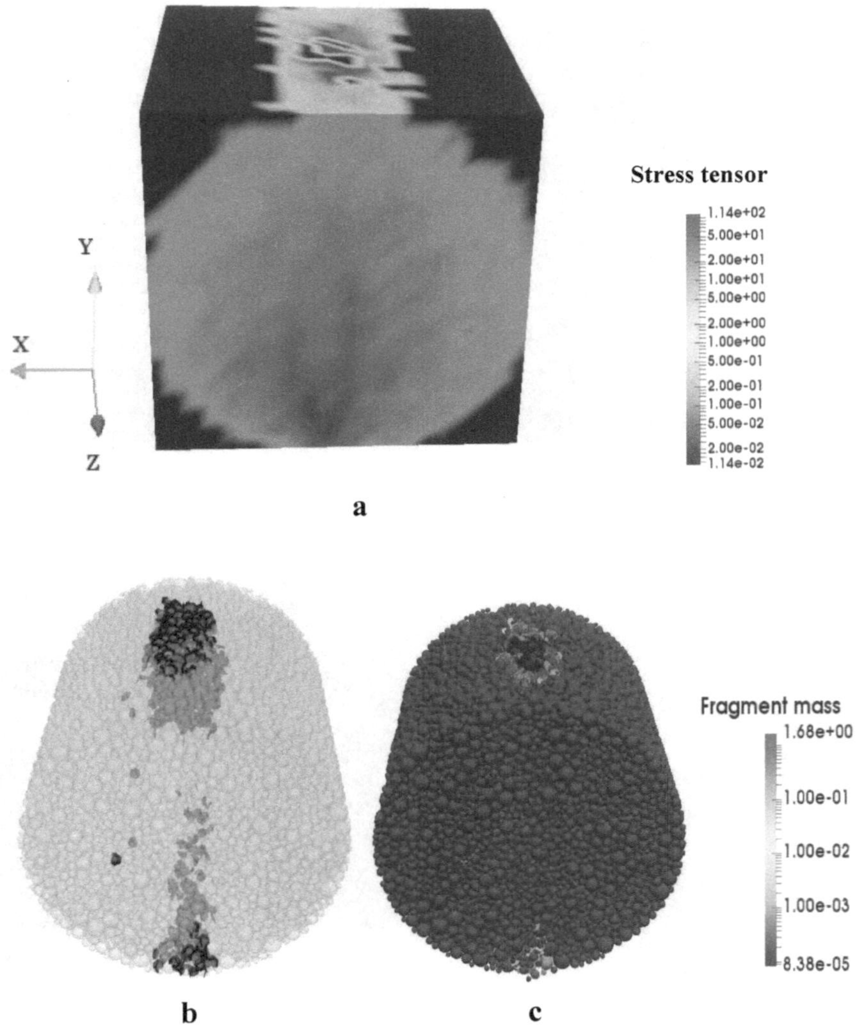

Fig. 4.7: Load response at 0.02ms. (a) Stress distribution   (b) Fracture locations within the rock specimen, with DEM-spheres, made slightly invisible for clear visualisation. (d) Fragment formation

4.3.3   During mechanical failure

The duration of mechanical failure occurs in zone C from the peak force until there is no longer a significant fracture at about 0.03ms. Two instances (0.025 and 0.03ms) depict the behaviour in this zone. Just after the peak force (at approximately 0.022ms), the rock specimen yields under load when the force is greater than the overall rock specimen strength and causes catastrophic damage. The rock specimen experiences its maximum stress intensity close to the peak point (Fig. 4.8a) due to the accumulated strain energy, which consequently causes it to yield when it can no longer bear the load. The rock specimen failure is indicated by the steep drop in force. While the stress intensity may decrease, the dissemination of residual forces (Fig. 4.9a) throughout the rock specimen continues to generate damage. The rate of increase in the percentage of broken bonds is highest over this interval, initially rapidly increasing during the maximum intensity phase before decreasing and beginning to level out as the specimen cleaves. This coincides with the steep drop in accumulated strain energy as the rock specimen undergoes release in stress from the breaking of bonds and formation of fragments (Figs 4.8b 4.8, c, 4.9 b and 4.9 c). Majority of the strain energy is converted to kinetic energy, as the generated segments which break from the specimen move from the impact zone.

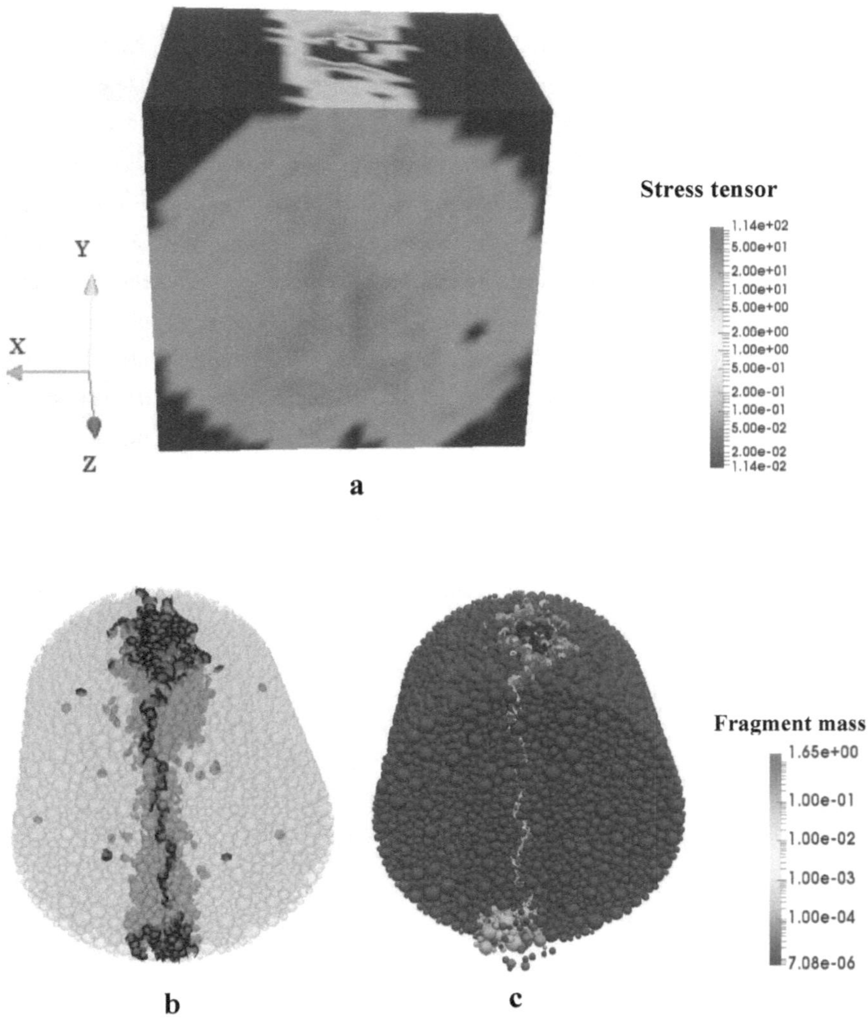

Fig. 4.8: Load response at 0.25ms. (a) Stress distribution viewed from the top. (b) Fracture locations within the rock, with DEM-spheres, made slightly invisible for clear visualisation. (c) Fragments cleave from the main body

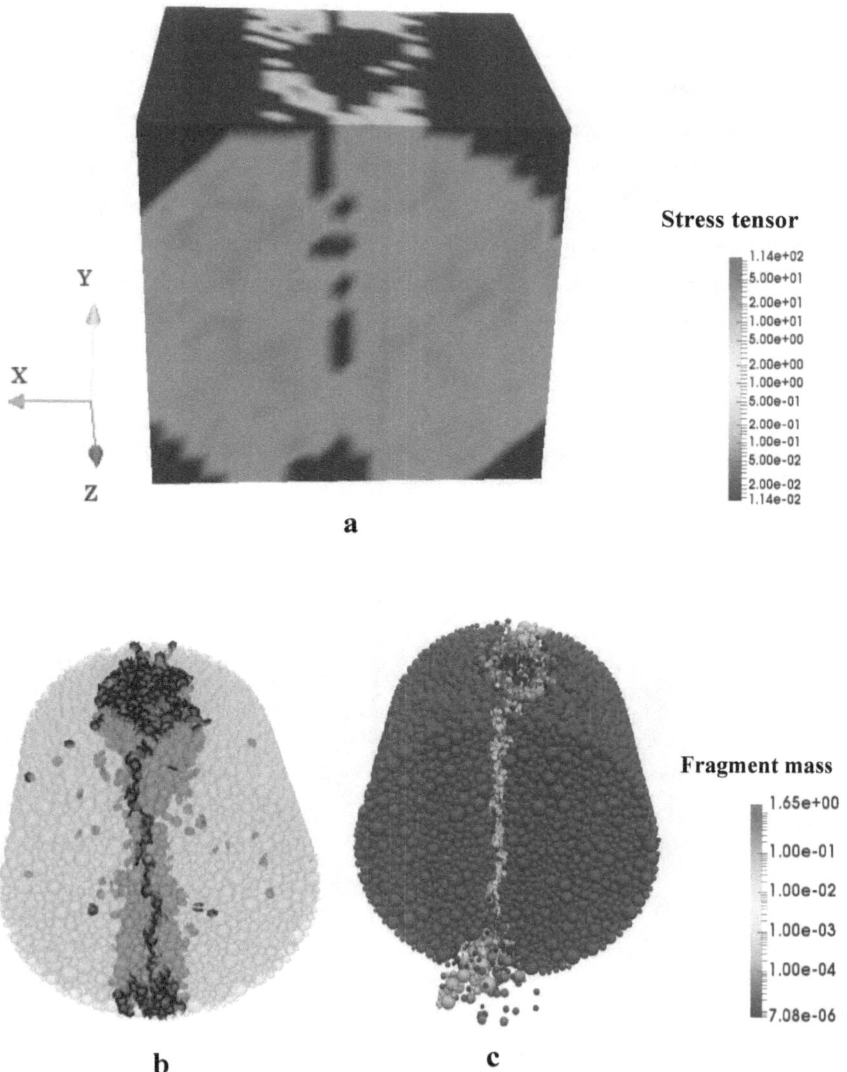

Fig. 4.9: Load response at 0.03ms. (a) Stress distribution viewed from the top (b) Fracture locations within the rock specimen, with DEM-spheres, made slightly invisible for clear visualisation. (c) Fragments cleave from the main body.

4.3.4   Subsequent to mechanical failure

This period (zone D) occurs after the major fracture has come to an end. The period is termed the post-fracture zone which occurs subsequent to failure and is depicted at 0.04ms in Fig. 4.10. The force has fully declined, and any remaining force spikes are of relatively low magnitude, caused mainly by contacting fragments which are still in motion away from the impact zone. This is further evidenced by the kinetic energy, which is at its highest when splinters form and escape the impact zone before levelling off at the end of the simulation with some fragments in motion. The accumulated strain energy approaches zero as the rock specimen is now under relatively little stress, and thus the percentage of broken bonds levels out.

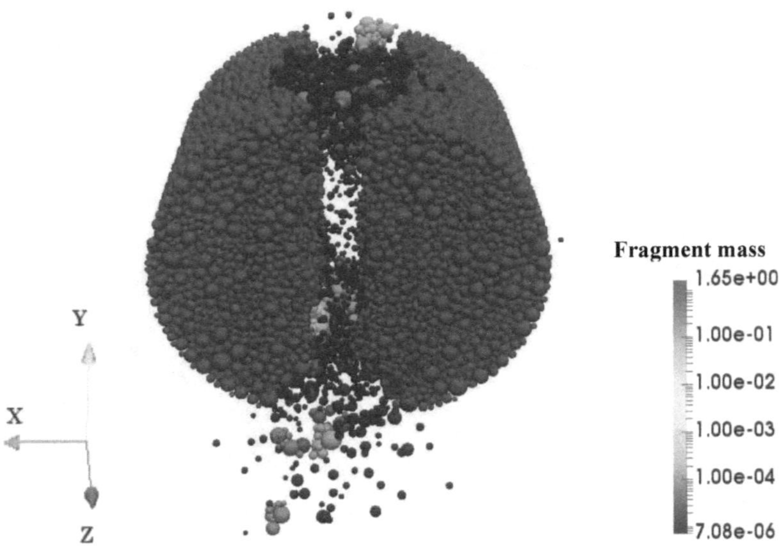

Fig. 4.10: Visualization of the rock specimen in the post-fracture zone (0.04ms)

4.4   Chapter summary

This chapter described the numerical methodology followed to create the rock specimen as well as the SILC testing environment. Data extracted from the DEM simulations were used to quantify and calculate pertinent fracture measures and establish consistent and defined analytical rules for quantifying these. The breakage behaviour during a single impact simulation was used to explain the key stages involved during an impact event viz: prior to impact, prior to mechanical failure or damage zone (stress build-up until it reaches peak), during mechanical failure or fracture zone (stress release and increased fracture locations) and subsequent to mechanical failure or post-fracture zone (fragmentation of rock specimen).

5 Evaluation of the bonded particle SILC model

This chapter is an excerpt from an article (Oladele et al., 2019) that has been submitted for publication and currently under review. A cubic-shaped specimen was used in the study and the comparison of the simulated breakage behaviour with the cylinder and conventional SILC results showed similar responses. This evidence is presented in Appendix I

The robustness of the current bonded particle SILC model was evaluated with respect to changes in model resolution, sample size and mechanical properties.

5.1 Model resolution sensitivity analysis

Model resolution has been demonstrated to have a significant effect on the accuracy of BPM simulations of rock damage (Charikinya, 2015). The effect of model resolution in this current study was examined by varying the scales of the DEM-spheres used to create rock specimens and measuring the variation in fracture force while keeping all other simulation parameters constant.

The model resolution is defined by a dimensionless ratio (Rmax/L) i.e., the ratio of the maximum DEM-sphere radius (Rmax) to the overall length of the specimen (L), where L is the same at all resolutions. By this definition, it is noteworthy that a low model resolution referred to a scenario with relatively few DEM-spheres while a high resolution indicated a relatively high number of DEM-spheres. Furthermore, because the length of the specimen is constant in this case, a high value of Rmax/L implies a low model resolution and vice-versa. The Rmax/L value was varied from 0.128 to 0.016 to test scenarios from low to high model resolution respectively, as demonstrated in Fig. 5.1. For each resolution, an equal ratio of the minimum radius of DEM-spheres (Rmin) to Rmax was maintained. This ensured the macroscopic mechanical properties of the rock specimens remained the same. Each simulation was repeated 30 times to provide adequate statistics for each scenario.

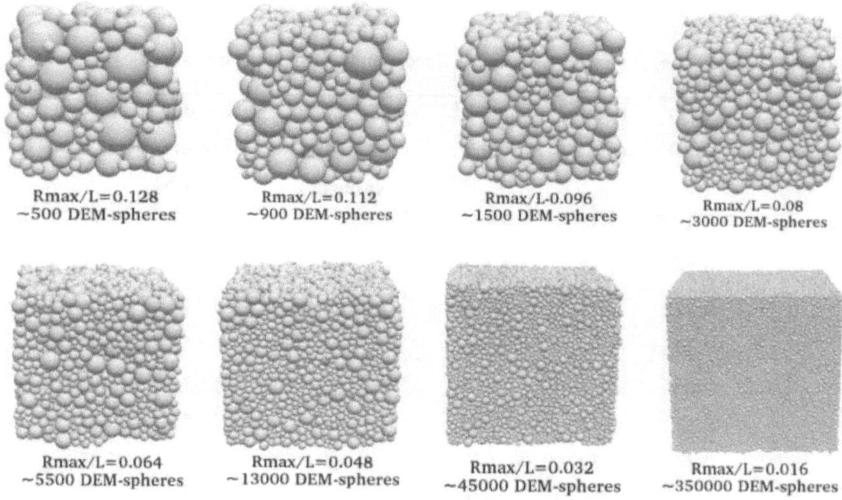

Fig. 5.1: A representative specimen at different model resolutions

Table 5.1 lists the parameters used for both *Rmin* and *Rmax* at different model resolutions as well as the approximate average number of DEM-spheres obtained from the numerical volume filling algorithm for constructing the specimens. In addition, the mean fracture force and standard deviation (SD) are also given. Fig. 5.2 shows a plot of the fracture force per specimen, showing the variation at different model resolutions. The average fracture force showed no statistically significant increase in the measured fracture. Notably, however, the fracture force converged toward more consistent values at high model resolutions (low values of Rmax/L). This implies that the measured fracture force becomes more accurate as the number of DEM-spheres used for simulations increases. Quantitatively, the standard deviation (SD) of the measured force decreased from approximately 1.56 to 0.18 between the highest and lowest Rmax/L values, highlighting that the variability of the measured fracture force is quite dependent on the number of DEM-spheres used to represent a rock specimen.

Table 5.1: Parameters for model resolution and measured force to fracture

Rmin (mm)	Rmax (mm)	Rmax/L	Approx. No. DEM-spheres	Average Fracture force (kN)	SD
0.32	1.28	0.128	550	5.15	1.56
0.28	1.12	0.112	900	4.65	1.49
0.24	0.96	0.096	1500	4.04	1.06
0.20	0.80	0.080	3000	3.84	0.77
0.16	0.64	0.064	5500	3.93	0.95
0.12	0.48	0.048	1300	3.65	0.59
0.08	0.32	0.032	45000	3.39	0.47
0.04	0.16	0.016	350000	3.04	0.18

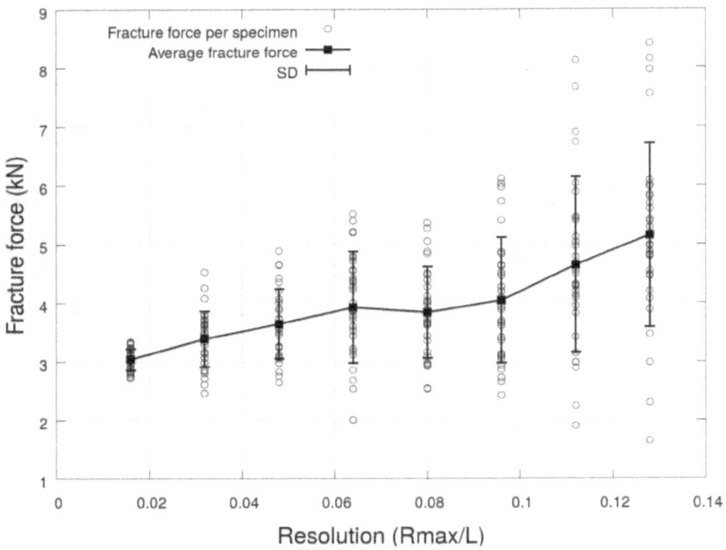

Fig. 5.2: Variation of fracture force with model resolution

The specific fracture energy and rock specimen stiffness were calculated according to the definitions given in Chapter 4. Customarily, the distribution of fracture energies from laboratory breakage tests conform to log-normal distributions when plotted cumulatively. For the DEM simulations, the three lowest values of Rmax/L (0.016, 0.032 and 0.064) were found to follow log-normal distribution models as shown in Fig. 5.3. Thereafter as this Rmax/L value increased beyond 0.096, the cumulative specific fracture showed progressively less of a tendency to follow this descriptor (Fig. 5.3). A realistic distribution could not be obtained at 0.128.

The mean and variation in mean specimen stiffness are shown in Fig. 5.4. The mean does not show a statistically significant change for all resolutions. However, its variation increases substantially from a Rmax/L value of 0.096 to 0.128.

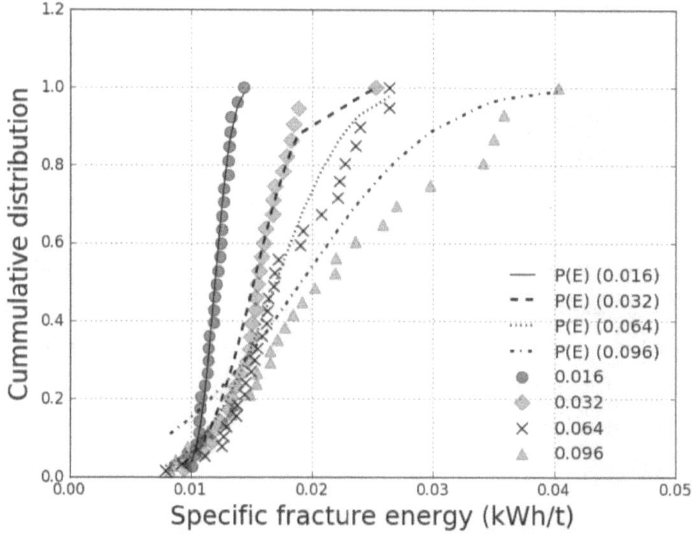

Fig. 5.3: Variation of fracture energy with model resolution. model resolutions at 0.016, 0.032 and 0.064 obeyed log-normal distribution fit., resolution at 0.096 fairly follows the log-normal distribution. A realistic distribution could not be obtained at 0.128. P(E) represents the log-normal distribution model fit

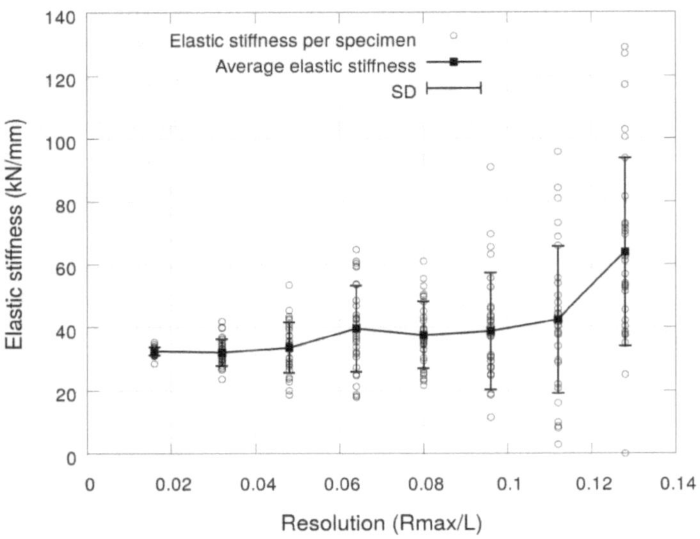

Fig. 5.4: Variation of elastic stiffness with model resolution

5.2 Sample-size dependency

The BPM sensitivity to measured fracture characteristics was also studied for different rock specimen sizes ranging from 2-10mm in length. Simulations for different specimen lengths were conducted at a resolution of 0.032 as the prior section shows that it provided a reasonable balance between its associated variability and computational expense. Simulations with this resolution ran for 15 hours while the higher resolution required approximately twice the computation time.

The fracture force plotted against size in Fig. 5.5 increased with specimen length. In addition to this is the plot of the elastic stiffness at different specimen length in Fig. 5.6. The elastic stiffness also showed a size-dependency with specimen length, with a more linear relationship. The elastic stiffness is an extrinsic property akin for the fracture force that has a direct relationship to the fracture energy. The elastic stiffness is different from "particle stiffness" or better termed "apparent elastic modulus" defined by Tavares and King (1998). This implies that its values scales with specimen size. Hence, the value of these properties increases as the specimen length increases.

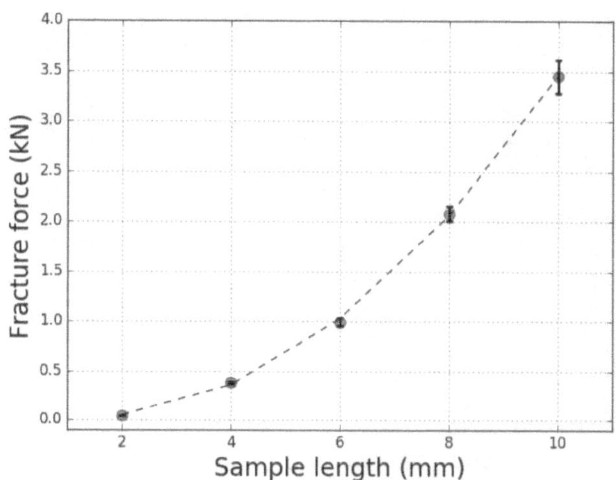

Fig. 5.5: Variation of fracture force with specimen length

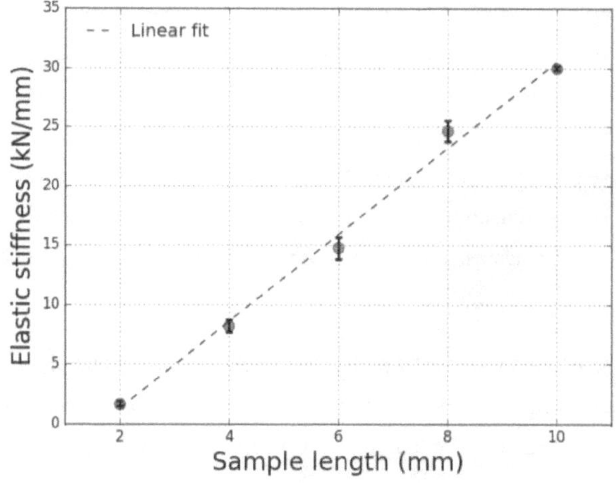

Fig. 5.6: Variation of elastic stiffness with specimen length

The cumulative distribution of the specific fracture energy for different specimen lengths was found to follow the log-normal distribution as shown in Fig. 5.7, for which 30 specimens were used for each specimen length.

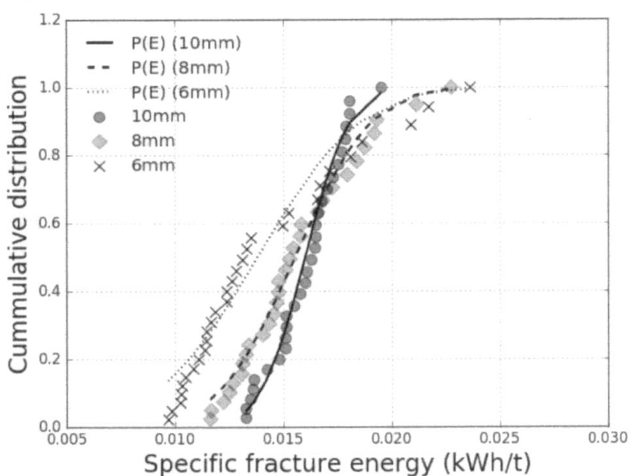

Fig. 5.7: Variation of specific fracture energy distribution at different specimen lengths.

5.3 Variation of mechanical properties

It is well-known that a material's Young's modulus (Y) and unconfined compressive strength (UCS) are important macro mechanical properties in rock mechanics and most often measured from loading experiments. While Y describes the elastic property of a rock specimen prior to failure, the UCS gives a measure of the brittle nature of the rock. Model parameters are usually calibrated against this set of macroscopic properties for breakage studies (Section 3). The sensitivity of the rock specimen fracture characteristics in a DEM-SILC setup was studied against these two mechanical properties at a model resolution of 0.032.

Figs 5.8 and 5.9 respectively show the plot of the fracture force against the UCS and Y that were obtained from SILC simulations. The UCS and Y which are a function of DEM beam interactions were systematically varied. Ten statistically identical specimens were employed for each UCS and Y combination for a total of 350 simulations. Fig. 5.8 shows that the fracture force under impact loading condition varies significantly with the UCS at a given Young's modulus. On the other hand, fracture force shows minor dependency on Young's modulus at a given UCS as shown in Fig. 5.9. Fig. 5.10 summarises the variation of Y and UCS against the fracture force in a 3D surface plot. This plot serves to demonstrate two important points. Firstly, the per-specimen variability of fracture force is significantly smaller than the range of fracture forces achievable by tuning the DEM beam parameters. This implies that the model results do not vary significantly for a given choice of model parameters/macroscopic properties at the DEM-sphere resolution employed herein. In

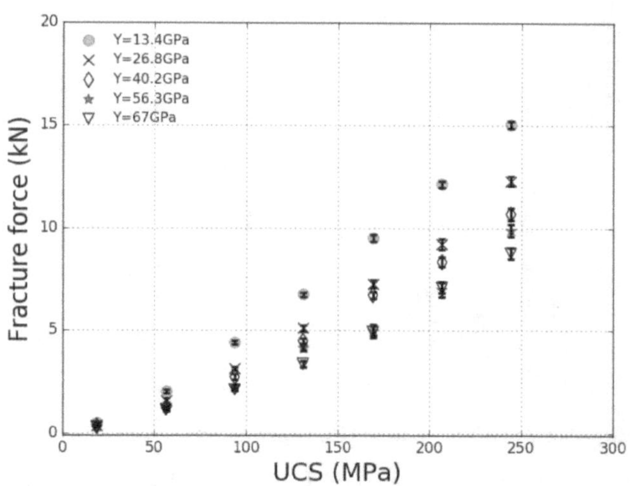

Fig. 5.8: Fracture force versus UCS at different Young's moduli

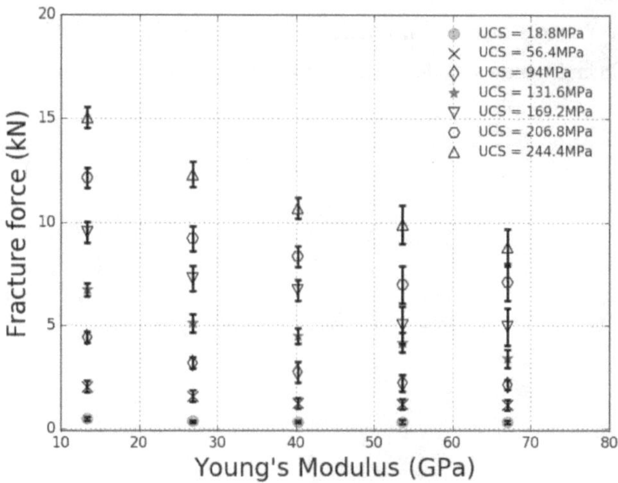

Fig. 5.9: Fracture force versus Young's modulus at different UCS

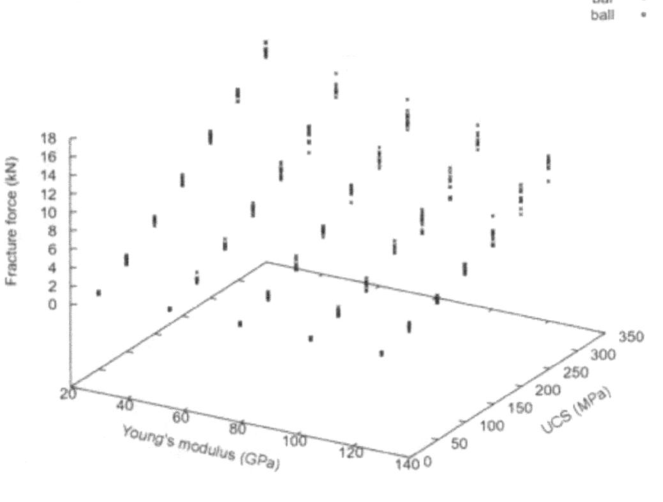

Fig. 5.10: Surface plot of Young's modulus and UCS against fracture force

Likewise, Figs. 5.11 and 5.12 show plots of the elastic stiffness against the UCS and the stiffness against Young's modulus respectively. It can be observed from these plots that the elastic stiffness is influenced by the UCS and Young's modulus. However, the UCS shows a statistically significant effect at different values of Young's modulus (Fig. 5.11) as compared to the effect of Young's modulus at different UCS (Fig. 5.12). Fig. 5.13 summarises the variation of UCS and Young's modulus against the elastic stiffness on a surface plot.

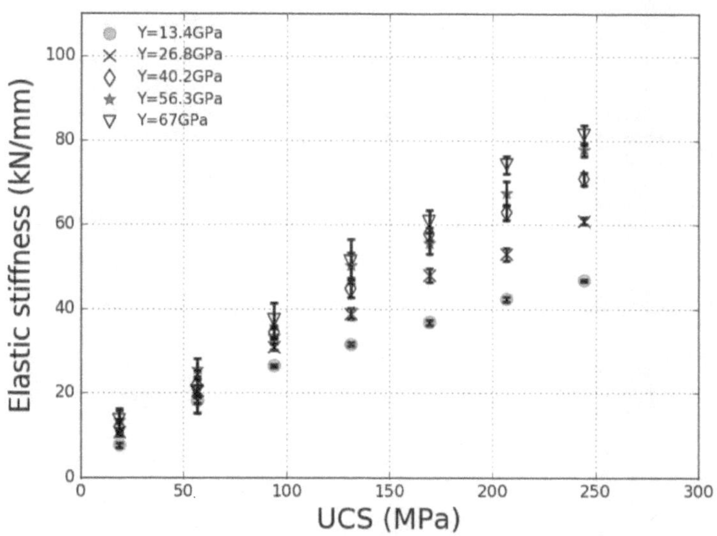

Fig. 5.11: Elastic stiffness versus UCS at different Young's modulus

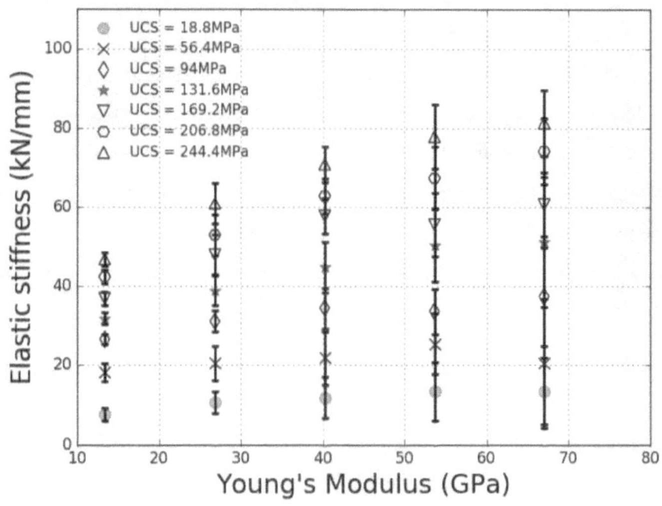

Fig. 5.12: Elastic stiffness versus Young's modulus at different UCS

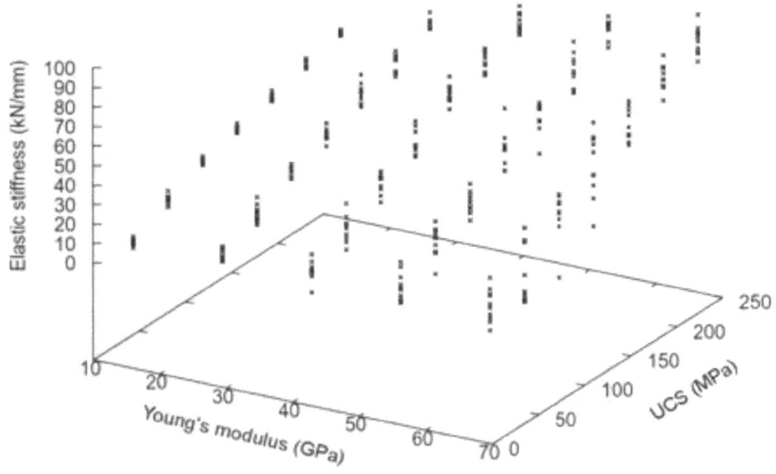

Fig. 5.13: Surface plot of Young's modulus and UCS against elastic stiffness.

Fig. 5.14 shows the plot of the specific fracture energy by the UCS at different values of Young's modulus while Fig. 5.15 presents the plot of this energy versus Young's modulus at different UCS. Summarily, the estimated specific fracture energy across these ranges of UCS and Young's modulus is also presented as a surface plot in Fig. 5.16. This also shows a dependency on both Young's modulus and UCS which increases against UCS but decreases with Young's Modulus. The cumulative distribution considered for two different Young's moduli and three UCS values are used to illustrate this dependency as shown in Fig. 5.17. These results suggest that the specific fracture energy is more likely to be influenced by the strength than Young's modulus, as a wider variation is noted for specific fracture energies at different strengths (at the same Young's modulus) compared to different Young's moduli (at the same strength).

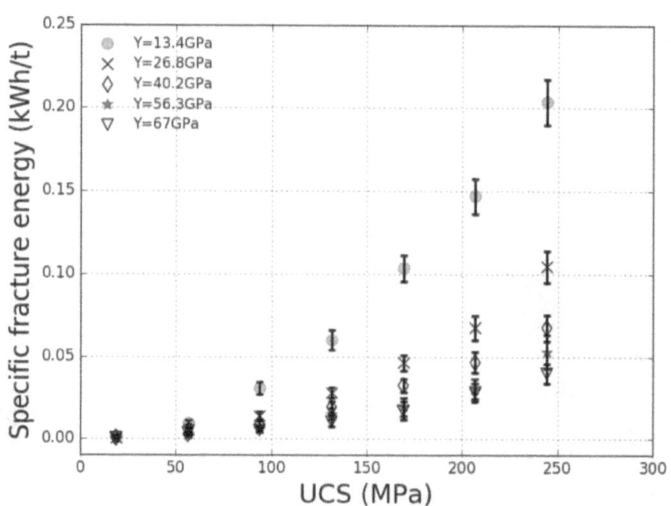

Fig. 5.14: Specific fracture energy versus UCS at different Young's moduli

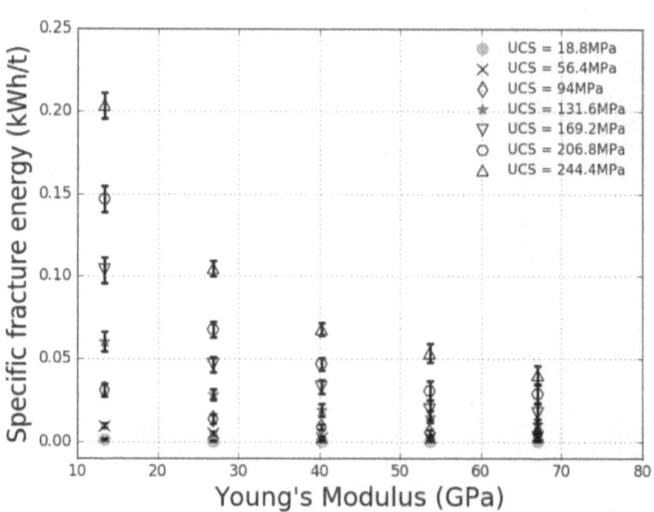

Fig. 5.15: Specific fracture energy versus Young's modulus at different UCS

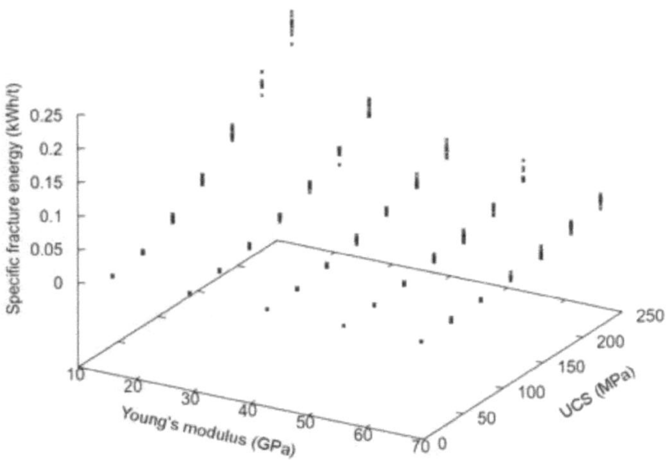

Fig. 5.16: Surface plot of Young's modulus and UCS against specific fracture energy

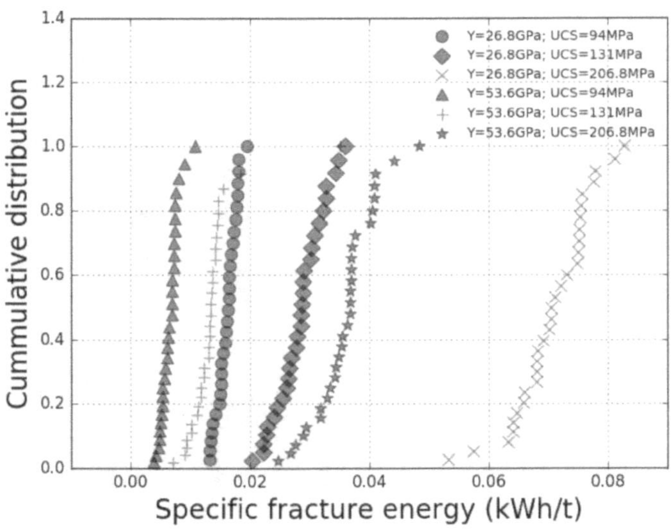

Fig. 5.17: Cumulative distribution of specific fracture energy at different combinations of Young's modulus and cohesive strength

5.4 Chapter summary

The BPM in DEM has been demonstrated to be a robust breakage model that provides an opportunity to study both intrinsic and extrinsic rock properties against the observed response during impact breakage. Three measures, namely; fracture force, stiffness and fracture energy were used to define the fracture characteristics. Model resolution, specimen size effect and variation of macro and microscopic mechanical properties were used and statistically analysed to check the sensitivity of the model against the fracture measures.

A model resolution with a sufficiently high number of DEM-spheres to reduce deviation and error to reasonable levels was shown to be important to achieve numerically stable results. The dependency of fracture characteristics on specimen size showed the expected behaviour of the homogeneous specimens considered in this study. The variation of Young's modulus and UCS against the fracture measures emphasised that the locus of these properties against the fracture measure can be used to suggest a calibration relationship.

6 A comparative study of DEM simulations with laboratory experiments

This chapter reports a comparative study of impact breakage of single rock specimen using the current BPM-DEM methodology against laboratory SILC experiments conducted by Barbosa et al. (2019). These authors conducted SILC experiments on synthetic rock specimens (artificial sandstones), produced via three-dimensional (3D) printing. The main advantage of having a synthetic rock specimen is the control of shape and size of rock specimens with minimal flaws (Johnston and Choi, 1986). This allows repeatable measurement of mechanical properties of such specimens (Johnston and Choi, 1986).

Barbosa et al. (2019) considered four different shapes of synthetic rock specimen which included; spheres, flattened-spheres, ellipsoids, and cylinders. Some of these shapes were printed in three different sizes, while others were printed in four different sizes. The aspect ratio (length: diameter) of all printed specimens used for the SILC experiments was approximately 1:1.

Numerical simulations in this work focused on cylinders and spheres. The empirical relationships developed in chapter 3 were applied to calibrate DEM model parameters to the macroscopic properties that were measured from experiments. The experimental methodology of Barbosa et al. (2019) and the DEM approach employed in this study are described in subsequent sections.

6.1 Experimental procedure

6.1.1 Manufacture of synthetic rock specimens

Synthetic rock specimens used in the study (Barbosa et al., 2019) were produced using a 3D-printing method (Section 2.5). 3D-printed sandstones were fabricated using jet binding technology and silica sands. All specimens were produced with 8% by volume (20% binder saturation) and post-treated at 80°C for 24 hours (Hodder, 2017). Specimens were printed in layers of sand grains and the thickness of each layer was approximately 250µm.

Two methods of printing were adopted for the 3D-printed cylinders; a vertical printing and horizontal printing direction as illustrated in Fig. 6.1 a and b respectively. The printing direction of spherical specimens could not be noticeably ascertained after completion. The loading direction for both vertically and horizontally printed cylinders was the same i.e., specimens were placed horizontally on the steel rod and impacted from the top as exemplified in Fig. 6.1c. The load response of the impact events was measured from the SILC device as previously described in section 2.3.3. Tables 6.1 to 6.3 summarise the dimensions of the synthetic rock specimens and the operating parameters of the SILC device (Barbosa et al., 2019).

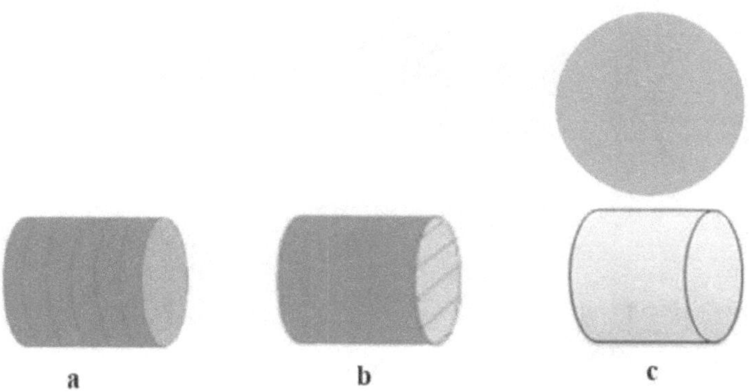

Fig. 6.1: An Illustration of printing direction (red lines showing printing direction in layers) and loading of the 3D-printed cylinders (Barbosa et al., 2019). (a) vertically printed cylinder (b) horizontally printed cylinder and (c) placement (horizontal) and loading direction for both vertical and horizontal printed cylinders

Table 6.1: Dimension of vertically 3D printed cylinder tested on SILC

Shape	Vertically printed cylinder			
Specimen size (mm)	Φ5.3	Φ7.38	Φ10.18	Φ12.36
Aspect ratio	1:0.98	1:1.04	1:0.98	1:0.99
Number of tests	21	26	26	16
Mass of impactor (g)	535.25	535.25	535.25	535.25
Drop height (mm)	25.07	63.33	100	125.11
Picture				

Table 6.2: Dimension of horizontally 3D printed cylinder tested on SILC

Shape	Horizontally printed cylinder		
Specimen size (mm)	Φ7.2	Φ10.01	Φ12.1
Aspect ratio	1:1.01	1:1.11	1:1.12
Number of tests	19	19	19
Mass of impactor (g)	535.25	535.25	535.25
Drop height (mm)	63.33	100	125.11
Picture			

Table 6.3: Dimension of 3D printed spheres tested on SILC

Shape	Sphere			
Specimen size (mm)	Φ5.1	Φ7.2	Φ10	Φ12.2
Number of tests	20	20	20	20
Mass of impactor (g)	535.25	535.25	535.25	535.25
Drop height (mm)	25.07	63.33	100	125.11
Picture				

6.1.2 Synthetic rock specimen mechanical properties

Uniaxial compression strength (UCS) tests were conducted on vertically printed cylinders using an axial extensometer to measure the stress and strain up to the peak strength (Barbosa et al., 2019). An example of the stress-strain curve obtained for the printed specimens is shown in Fig. 6.2 (Hodder, 2017; Barbosa et al., 2019). The Young's modulus of the material was calculated from the slope of the linear portion of the stress-strain curve and UCS was estimated from the peak stress. The Young's modulus and UCS were characterised for each size of the synthetic rock. Table 6.4 summarises the results of Young's modulus and UCS obtained from the tests (Barbosa et al., 2019)

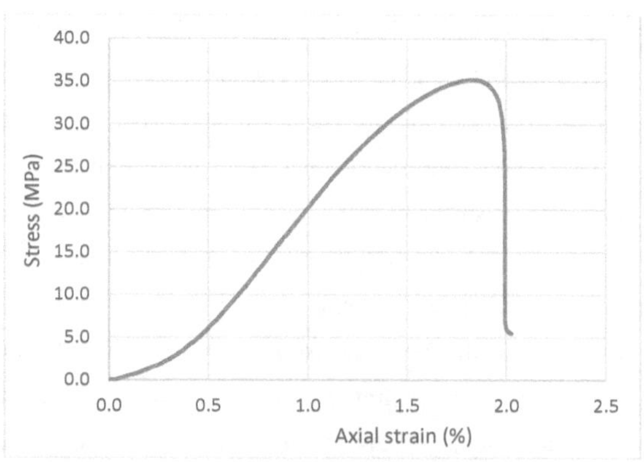

Fig. 6.2: Typical stress-strain curve obtained for 3D-printed rock specimens (Hodder, 2017; Barbosa et al., 2019)

Table 6.4: Summary of the UCS and Young's modulus measured for five specimen sizes (Barbosa et al., 2019)

Sample diameter (mm)	7.3	10.1	12.3	16.5	21.6
UCS (MPa)	20.1	24.5	27.9	32.5	34.7
Y (GPa)	1.3	1.8	2.0	1.9	1.9

6.2 Numerical procedure and comparative results

Two main steps were taken for the numerical procedure before conducting the comparative study. The first was to calibrate the model parameters against the measured macroscopic mechanical properties (Young's modulus and UCS) that were obtained from experiments. The second involved setting up the numerical SILC setup and running the simulation.

6.2.1 Calibration of model parameters

It was shown in chapter 3 that the macroscopic Young's modulus is significantly dependent on the beam elastic modulus (Y_b) compared to the beam Poisson's ratio (v_b) under compression and tension. Also, the macroscopic UCS is significantly influenced by the beam cohesive strength (C_b) than the internal angle of friction (θ_b). Hence, relationships were developed and established to relate model parameters with these macroscopic properties (Section 3.3). Furthermore, the goodness of fit with the R-square value also suggested that these relationships are quite valid. Therefore, rather than conducting a suite of UCS simulations for a range of model parameters and measuring the macroscopic Young's moduli and UCS values to match the measured values from the experiment, the applicable set of relationships with determined empirical constants were used to calculate the model parameters at the measured experimental UCS and Young's modulus. Eqs. 3.1 and 3.3 were found to be applicable to determine the corresponding model parameters. These are equations of Young's modulus under compressive loading and UCS also shown in Eqs. 6.1 and 6.2. Reasonable default values for the beam Poisson's ratio ($v_b = 0.25$) and internal angle of friction ($\theta = 45°$) were assumed. Substituting these values into the calibration relations (Eqs 6.1 and 6.2) permitted calculation of appropriate values for the beam elastic modulus and cohesive strength. Table 6.5 and 6.6 list the calculated and chosssen values that were employed respectively.

$$Y_c = \frac{0.568\, Y_b}{(1+0.236\, v_b)} \qquad\qquad 6.1$$

$$UCS = \frac{1.85 C_b}{1.46 + \tan\theta} \qquad\qquad 6.2$$

Table 6.5: Values of the macroscopic parameters used to determine model parameters

Macroscopic parameter	Average value	SD
Young's modulus	1.8 GPa	0.23
UCS	27.9 MPa	5.29

Table 6.6: Model parameters

Model parameter	Average value	Comment
Y_b	3.35 GPa	Calculated from equation 6.1
C_b	37.2 MPa	Calculated from equation 6.2
θ	45 °	Chose typical value
v_b	0.25	Chose typical value

6.2.2 Numerical SILC setup

The numerical SILC setup was carried out in a similar manner to the description in chapter 3 except for applying the volume filling algorithm to cylindrical and spherical geometries. Figs

6.3a and b show an illustration of geometrical specimens and numerical SILC setup for both cylinder and sphere respectively. The corresponding impact velocity with respect to the drop height was calculated and prescribed for the impactor accordingly for each simulation scenario.

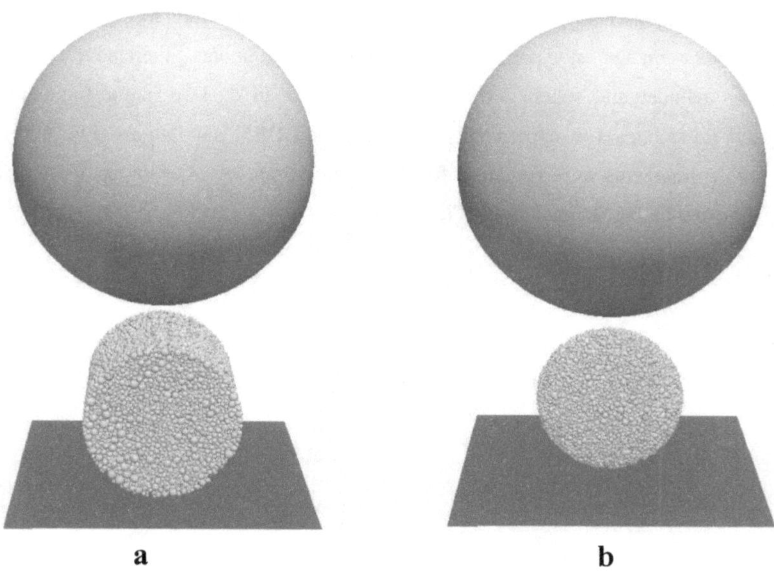

Fig. 6.3: An illustration of the numerical SILC setup of the 3D printed cylinder (a) and sphere (b).

6.3 Comparative analysis and results

Having "calibrated" model parameters using the fitting model against measured macroscopic Young's modulus and UCS obtained from experiment and calculated the corresponding numerical model parameters, DEM size dependency tests were conducted. These were then compared to SILC experiments.

Due to the different printing modes (vertical and horizontal) of the cylindrical specimens that were adopted in the experiment, it is not known which specimens correspond to the cylindrical DEM rock specimen. As such, the size dependency tests obtained for both vertically and horizontally printed cylinders were both compared against the DEM cylinders.

Fig. 6.4 presents the result of the size dependency test for the vertically and horizontally printed cylinders and spherical specimens compared against DEM cylinders and spheres. Results show that there is an increase in fracture force as the size of the rock specimen increases for all shapes. This is a consistent prediction of the experimental fracture force by the DEM simulations, particularly for the cylinders. Although the prediction from the spherical specimens also gives an increasing trend, DEM fracture force shows a slightly reduced prediction against experiment. Furthermore, the fragments generated after the impact event for cylinder and sphere are shown in Fig. 6.5a to d. Two major fragments were produced for cylinders in experiments (Fig. 6.5a) and DEM simulations (Fig. 6.5b) while three major fragments were produced for the spheres from the experiments (Fig. 6.5c) and complemented by DEM simulations (Fig. 6.5d)

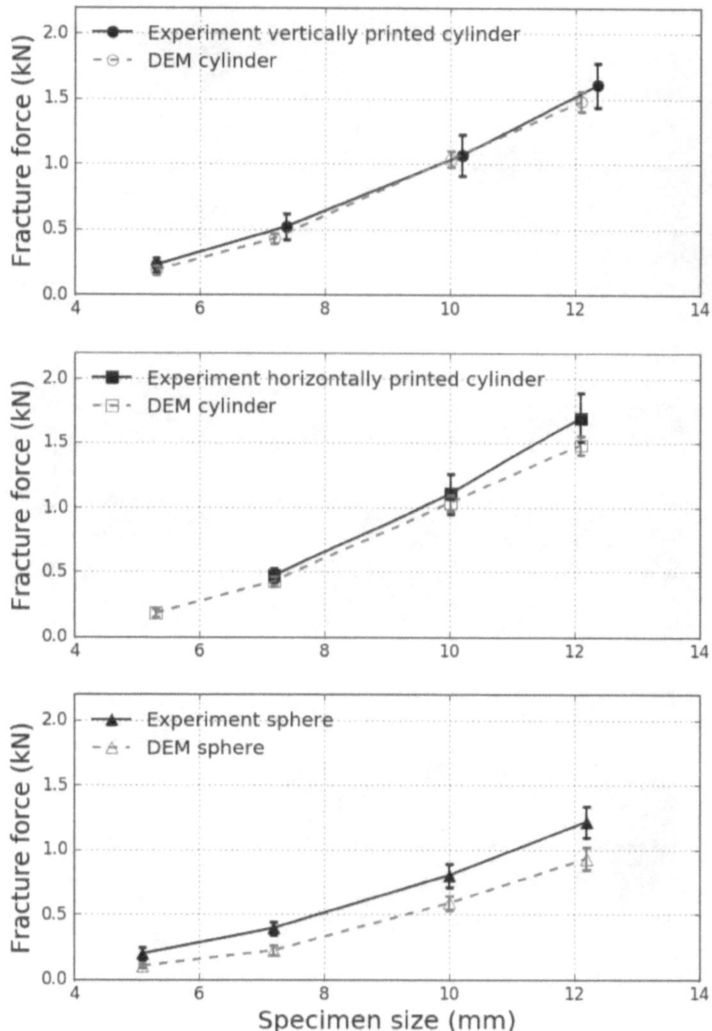

Fig. 6.4 (a) Fracture force measured for different sizes for the vertically printed cylinder compared against DEM-cylinders, (b) horizontally printed cylinder compared against DEM cylinders, and (c) sphere experiment compared against DEM spheres.

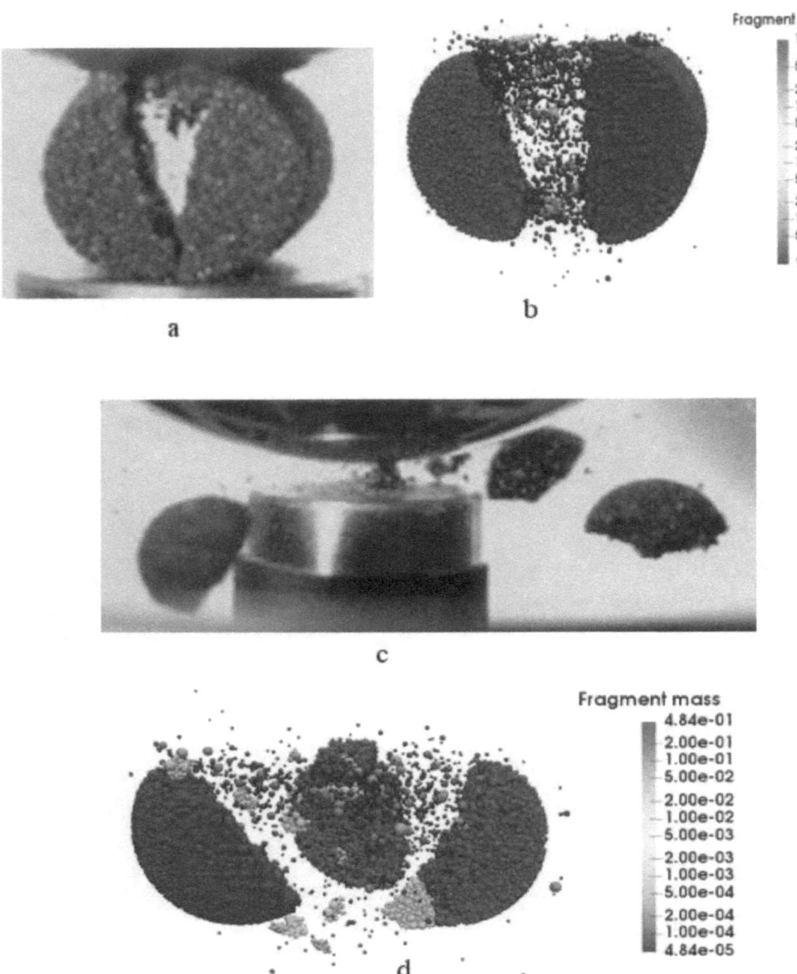

Fig. 6.5: Fragments generated from cylinders (a and b) and spheres (c and d). Photos 6.5a and c were extracted from the work of Barbosa et al. (2019).

6.4 Chapter summary

This chapter presents a comparison of the impact breakage of synthetic rock specimens using a short impact load cell (SILC) experiments against discrete element method (DEM) numerical simulations. The findings show that DEM simulation consistently predicts the fracture force obtained from the experiment for both cylindrical and spherical 3D printed synthetic rock specimens. This indicates that DEM can further be used as a complementary tool to investigate the effect of other factors which would be difficult to replicate using experiments that affect breakage behaviour.

7 Numerical investigations on pre-weakening and mineralogical composition

The results in chapter 6 have shown that the current BPM-DEM model can consistently predict the breakage of rock specimens reasonably well when compared against SILC experiments. Thus, further investigations were conducted on cylindrical "DEM-synthetic" numerical rock specimens to measure load response and study the effect of pre-weakening and mineralogical compositions within the rock specimen. For these investigations, all simulation conditions were kept consistent with the comparative study, except for the introduction of varying levels of pre-existing cracks and mineralogical compositions into synthetic rock specimens.

7.1 Investigation on pre-existing cracks

In practical comminution scenarios, most ores exist pre-weakened or pre-damaged with flaws and/or pre-existing cracks either from in-situ or during prior handling stages or both combined (Hahne et al. 2003). Two main methods of introducing pre-damage in rock specimens with the current BPM approach have been identified, which both produce similar results (William, 2014). The first method referred to as the micro-damage approach randomly removes bonds between discrete entities. The percentage of bonds removed is usually proportional to the level of pre-damage in the rock specimen. The second method known as the micro-crack method, represent cracks as rectangular planes or triangulated surfaces which are randomly distributed within the specimen. Discrete entities on either side of the planes interact via unbounded frictional interaction within the specimen geometry. An additional advantage that the micro-crack method has over the micro-damage method is that it captures the representation of damage in in-situ wherein planes of fracture tend to be semi-parallel to one another in different directions. In other words, the micro-crack method geometrically prescribes planes along which sliding tends to occur in addition to the flaws from prior handling processing in the comminution stages that contribute to the state of the strength of the rock specimen.

7.1.1 Construction of pre-weakened DEM rock specimen

Based on the advantages that have been highlighted in the previous section, the micro-crack method was employed in this study and pre-damage levels were quantified by the number of micro-cracks present in the rock specimen. It was found from preliminary geometry construction that the presence of the micro-cracks results in some discontinuities of the connecting bonds among the discrete entities, similar to the micro-damage method. An illustration is shown in Figs. 7.1a-f, where Figs 7.1a and b show the homogeneous specimen without micro-cracks and its bond networks respectively, while Fig. 7.1c shows a thin slice section from the bond network around the middle of the specimen in the Z-direction. Figs 7.1d and e show the specimen with micro-cracks and its bond networks respectively while Fig. 7.1f shows a thin slice section from the bond network around the middle of the specimen in the Z-direction. The presence of some of the cracks denoted by the disconnection between bond networks can be clearly seen in the thin slice section within the red circular rings in Fig 7.1f.

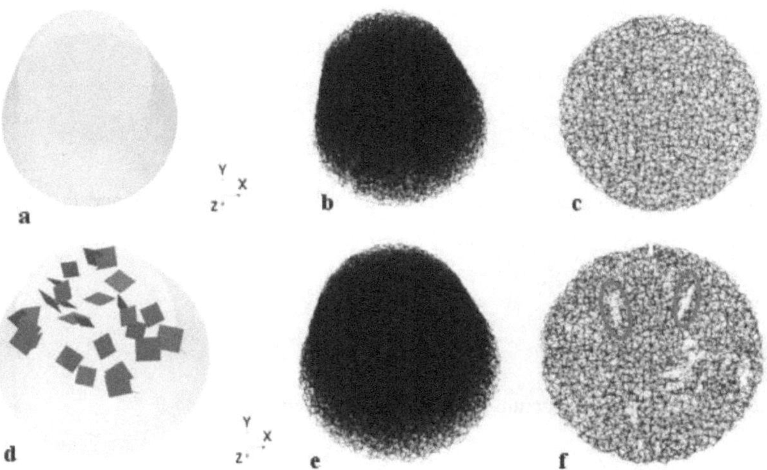

Fig. 7.1: (a) Homogeneous specimen without micro-cracks. (b) the bond network of the homogeneous specimen (c) a slice-cut of the bond network of the homogeneous specimen (d) specimen with micro-cracks. (e) the bond network of the specimen with micro-cracks (f) a slice-cut of the bond network of the specimen with micro-cracks

In view of this, the percentage level of pre-damage was estimated by working out the percentage ratio of the number of bonds in the presence of micro-cracks (N_{BP}) to the number of bonds in an equal volume of a homogeneous rock specimen (N_{BH}) as expressed in Eq. 7.1.

$$D_a = \frac{N_{BH} - N_{BP}}{N_{BH}} \; X \; 100\% \qquad\qquad 7.1$$

It is assumed that a homogeneous specimen has no pre-existing crack therefore, $D_a = 0$ and therefore taken as the basis with 100 percent unbroken bonds.

It is also important to differentiate between the fracture surface area and the new fracture surface area in this context and how they have been calculated. The fracture surface area is the surface area in the presence of a pre-existing crack before impact and was estimated according to Eq.7.2

$$FSA = \left(\frac{L_{min} + L_{max}}{2}\right)^2 Nc \qquad\qquad 7.2$$

Where, FSA = fracture surface area (mm^2)

L_{min} = minimum length of crack (mm)

L_{max} = maximum length of crack (mm)

Nc = number of cracks.

The new fracture surface area was estimated as the sum of the FSA and the surface area generated upon the breakage of the broken bonds between adjacent DEM-spheres (after impact). This is given by Eq. 7.3

$$NFSA = FSA + \pi \left(\frac{r_1 + r_2}{2}\right)^2 \qquad\qquad 7.3$$

Where $NFSA$ is the new fracture surface area generated (mm^2)

r_1 and r_2 are radii of two adjacent DEM-spheres on which the connecting bond is broken (mm)

7.1.2 Dependence of load response upon pre-weakening

Four levels (2, 4, 7 and 11%) of pre-damaged rock specimens with pre-existing cracks in a 10mm cylindrical DEM specimens were investigated with 30 repeats per level of damage.

Table 7.1 summarises the number of pre-existing cracks, the corresponding percentage of damage as well as the fracture surface area before impact that were investigated.

Table 7.1: Crack numbers and the corresponding percentage of damage before impact

Number of cracks	0	30	90	180	300
Da (%)	0	2	4	7	11
Fracture surface area (mm²)	0	59	176	353	588

Fig. 7.2 shows the plot of the measured fracture force at different percentages of pre-damage. The fracture force showed a decreasing trend with an increasing percentage of pre-damage. Homogeneous specimens without any pre-existing crack (0%) have the highest force to fracture of about 1.2kN and specimens with the highest percentage (11%) of pre-damage has the lowest force to fracture of around 0.6kN. The decreasing trend thus emphasized that the more damage accumulated, the weaker the specimen became, with the fracture force notably decreasing as the crack density increased beyond 4%.

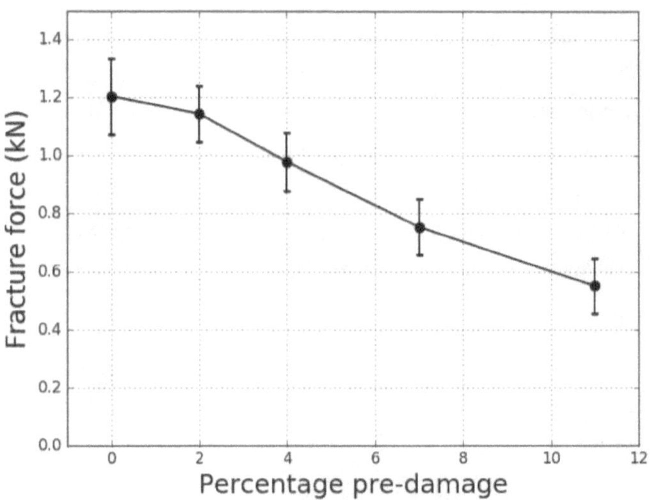

Fig. 7.2: Effect of pre-damage levels on the mean fracture force of 30 specimens. Error bar (SD) represents the standard deviation

Similarly, Fig. 7.3 shows the fracture surface area for initially a homogeneous rock specimen and then with increasing pre-existing cracks. Again, the measured fracture force decreases from around 1.2kN at 0mm^2 fracture surface area to about 0.6kN at about 580mm^2 fracture surface area.

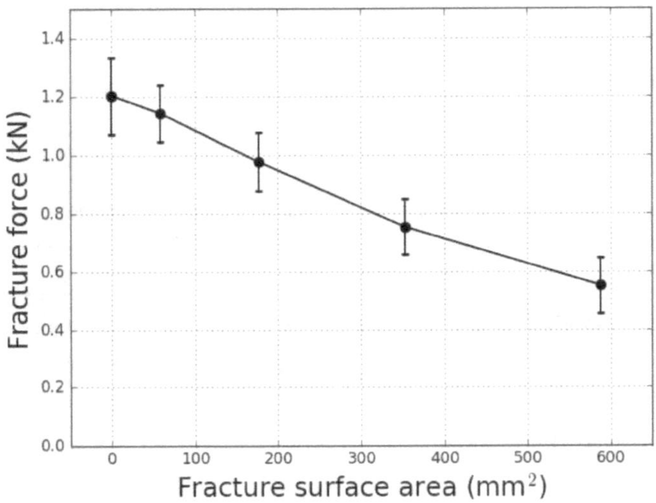

Fig. 7.3: Effect of fracture surface area on the mean fracture force. Error bar (SD) represents the standard deviation

Further analysis of the new fracture surface areas obtained after impact for these specimens is shown in Fig. 7.4. Results showed that rock specimens with the highest level of pre-damage (300 cracks) required the lowest force to fracture which also produced the least new fracture surface area. This trend towards a linear descriptor for the new fracture surface area as the number of pre-existing cracks decreases to zero.

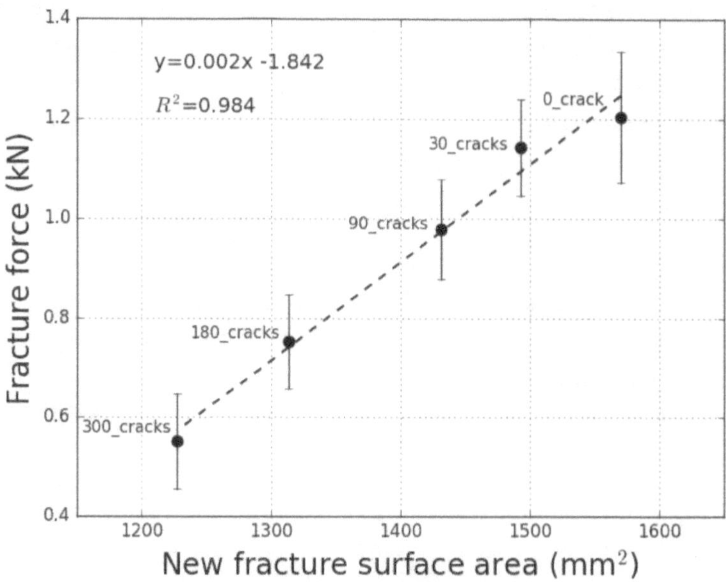

Fig. 7.4: Fracture force versus new fracture surface area

7.1.3   Breakage behaviour

Additional investigations were conducted on three random specimens at each pre-existing crack level to study the fracture behaviour. Results of the force-time histories, fracture pattern as well as the fragment size distributions are presented from Figs 7.5 to 7.10.

An examination of the force-time histories in these Figs indicate that homogeneous rock specimens with 0 cracks (Fig. 7.5a) and specimens with the least amount pre-existing cracks i.e., 30 cracks (Fig. 7.6a) exhibit the highest consistency of force-time histories. Histories are fairly consistent in the presence of 90 and 180 cracks as shown in Figs 7.7a and 7.8a respectively, while in the presence of 300 cracks (Fig. 7.9a), the force-time history shows more variability.

Furthermore, the fracture patterns leading to potential fragmentations for these three specimens at each crack level are shown in Figs 7.5 b, c, d to 7.9 b, c, d. The key observation here is also the consistency of the fracture pattern obtained at 0 crack level with specimens showing a tendency of splitting into almost two equal fragments (Figs 7.5 b, c and d). A similar split of rock specimens into two fragments was observed in the presence of 30 cracks into the rock specimen, where some inconsistencies in the symmetry of the cleavage begin to appear. This can be seen with fracture line of specimen 1 (7.6 b) moving slightly more diagonally to the left-hand side whereas the fracture lines of specimens 2 and 3 tend to pass through the centre of the specimen (Figs 7.6c and d). This behaviour can be attributed to the presence of micro-cracks driving the failure of crack propagation in variable directions within the rock specimen.

As the crack density intensifies, the irregularity of the fracture patterns within rock specimens become more apparent. These are shown with the fracture patterns at 90 cracks (Figs 7.7 b, c and d), 180 cracks (Figs 7.8 b, c and d) and 300 cracks (Figs 7.9 b, c and d). Among these, fracture patterns at 180 cracks and 300 cracks show the highest irregularities.

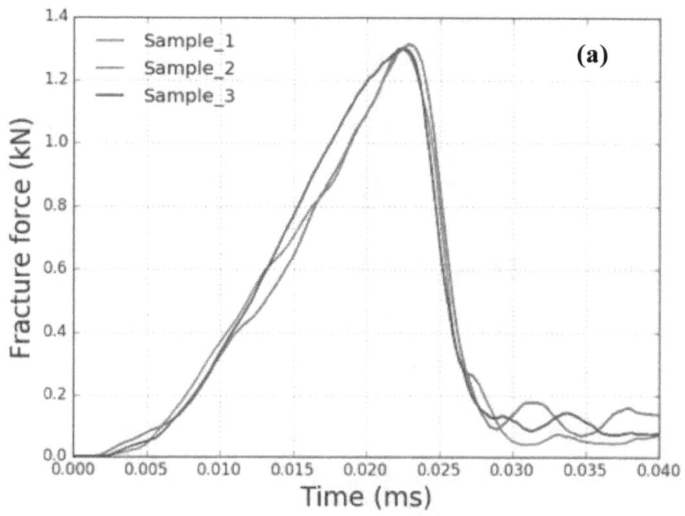

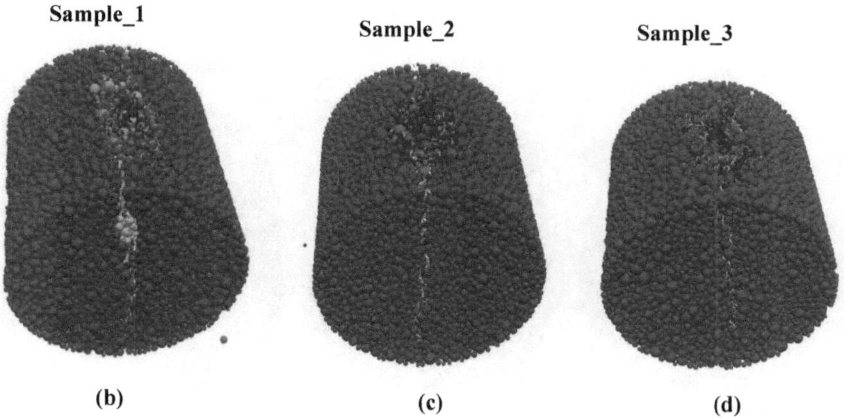

Fig. 7.5: Force-time histories of three rock specimens and their respective post-process analyses showing the fracture pattern at 0 crack level

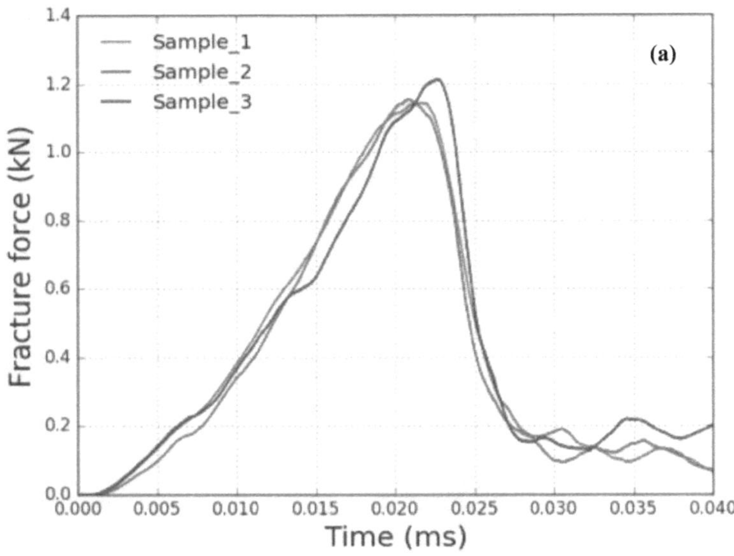

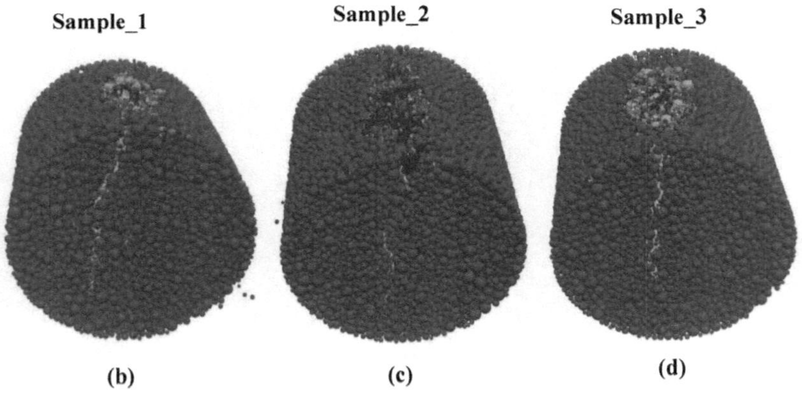

Fig. 7.6: Force-time histories of three rock specimens and their respective post-process analyses showing the fracture pattern at 30-crack level

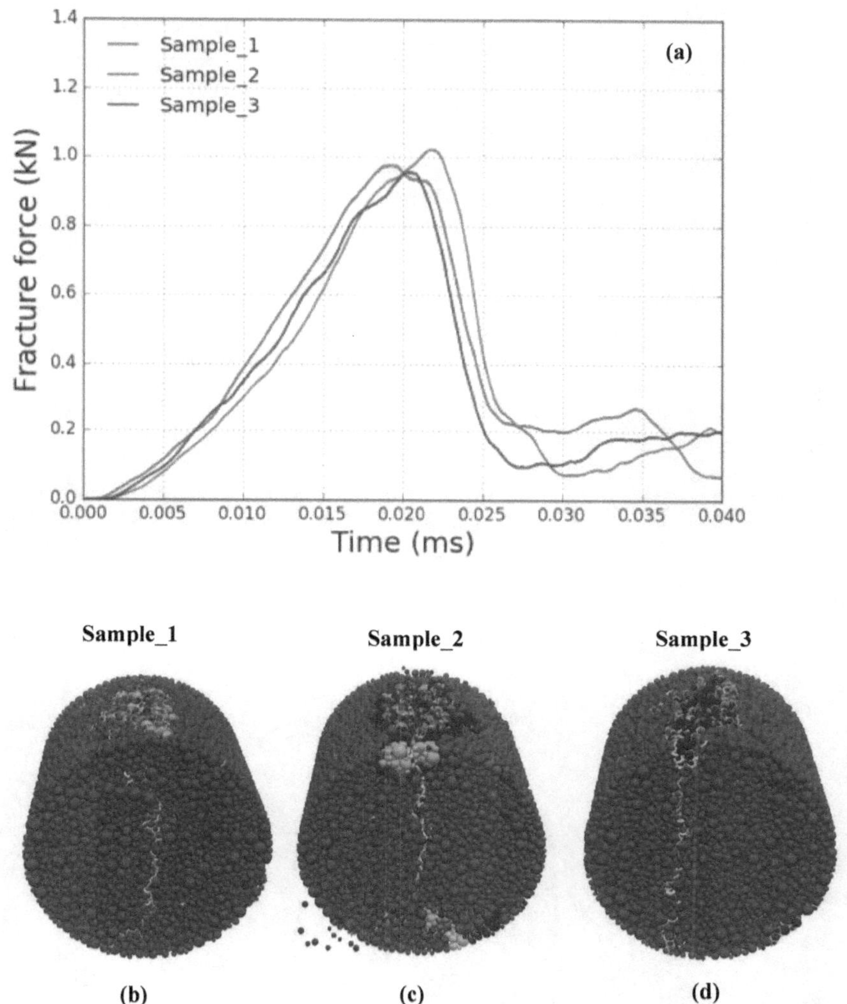

Fig. 7.7: Force-time histories of three rock specimens and their respective post-process analyses showing the fracture pattern at 90-crack level

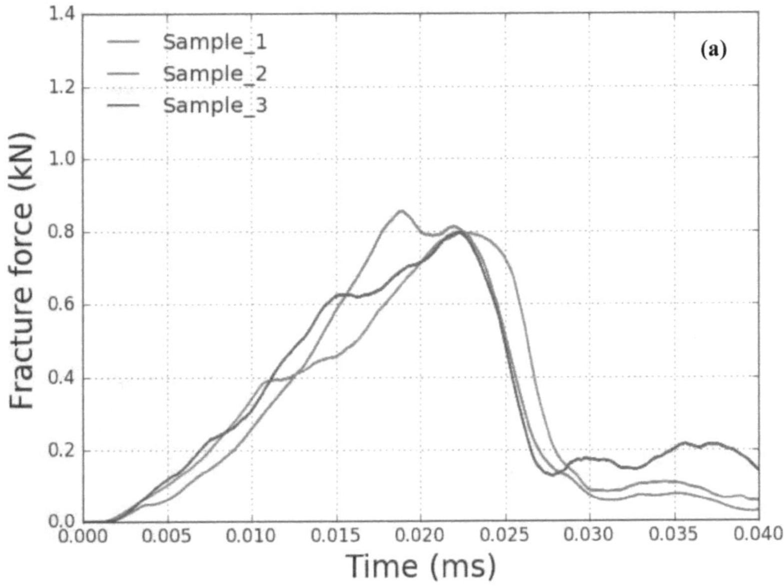

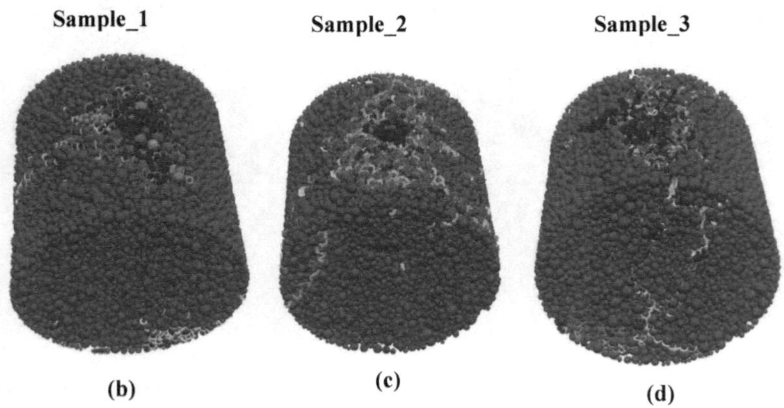

Fig. 7.8: Force-time histories of three rock specimens and their respective post-process analyses showing the fracture pattern at 180-crack level

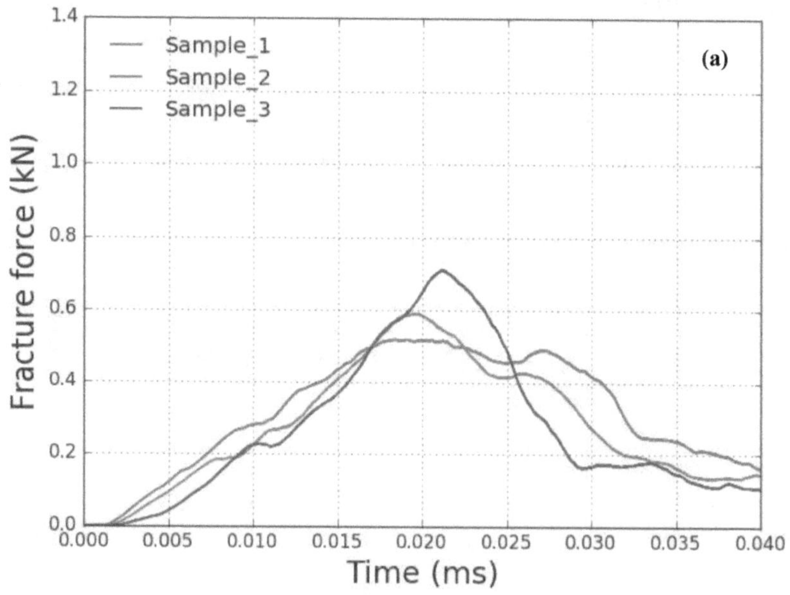

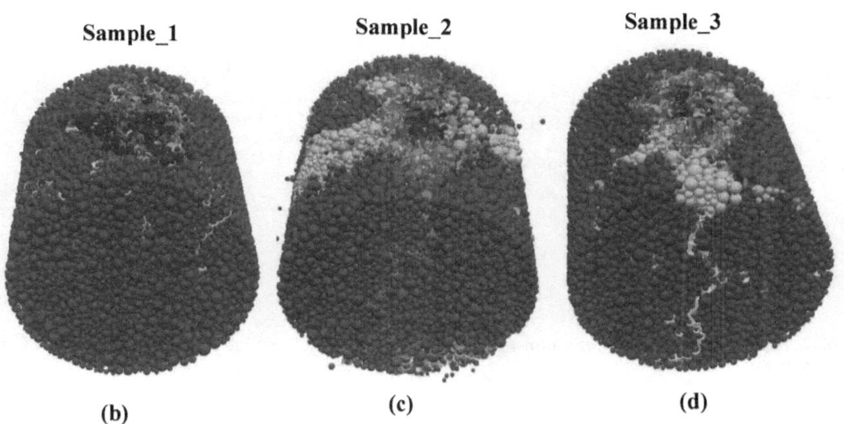

Fig. 7.9: Force-time histories of three rock specimens and their respective post-process analyses showing the fracture pattern at 300 crack level

The fragment size distribution (FSD) presented in Fig. 7.10 also complements the observations from the fracture patterns. This figure is divided into three sections; below 20% (fine fragments), 20-70% (medium size fragments) and above 70% (big fragments) cumulative passing. The importance of this figure is to show that only specimens with high crack densities (90, 180 and 300 cracks) produced medium size fragments according to the FSD. Fine fragments were produced in all cases as a result of the impactor crushing on the rock specimen at impactor-rock and rock-wall contacts. In addition, a fragment colour code of fragment masses at each crack level is shown in Figs 7.11a-e

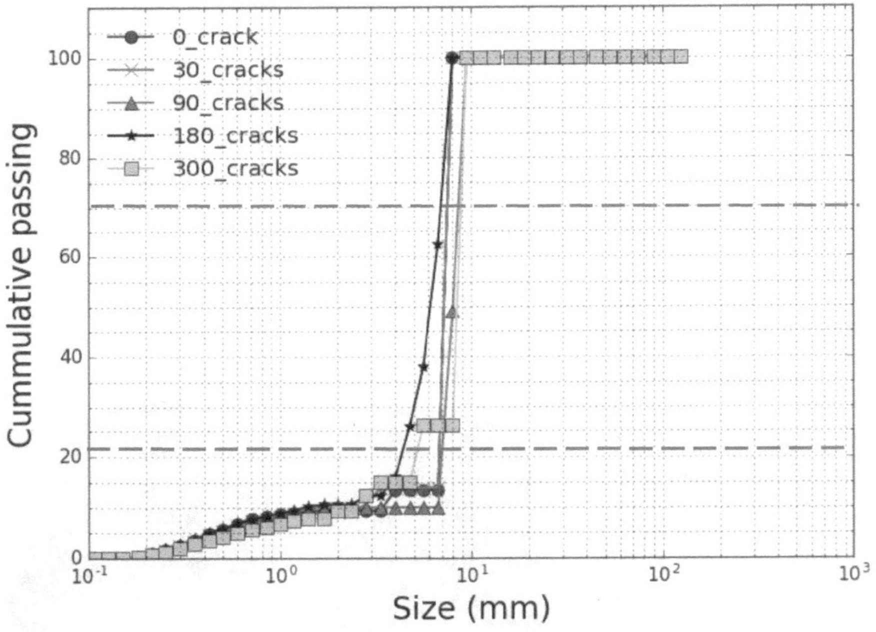

Fig. 7.10: Fragment size distribution of a representative rock specimen at different levels of pre-existing cracks

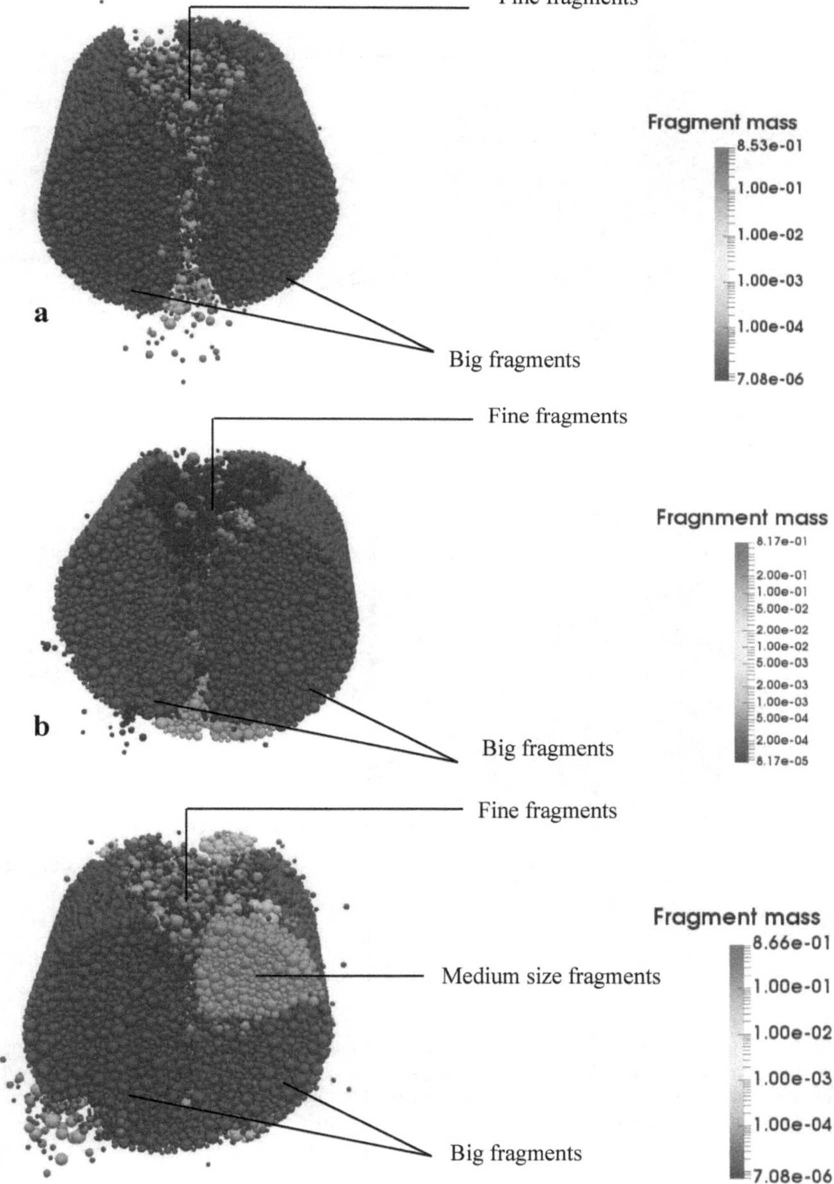

Fine fragments
Fragment mass
8.53e-01
1.00e-01
1.00e-02
1.00e-03
1.00e-04
7.08e-06
Big fragments
a
Fine fragments
Fragnment mass
8.17e-01
2.00e-01
1.00e-01
5.00e-02
2.00e-02
1.00e-02
5.00e-03
2.00e-03
1.00e-03
5.00e-04
2.00e-04
8.17e-05
Big fragments
b
Fine fragments
Medium size fragments
Fragment mass
8.66e-01
1.00e-01
1.00e-02
1.00e-03
1.00e-04
7.08e-06
Big fragments
c

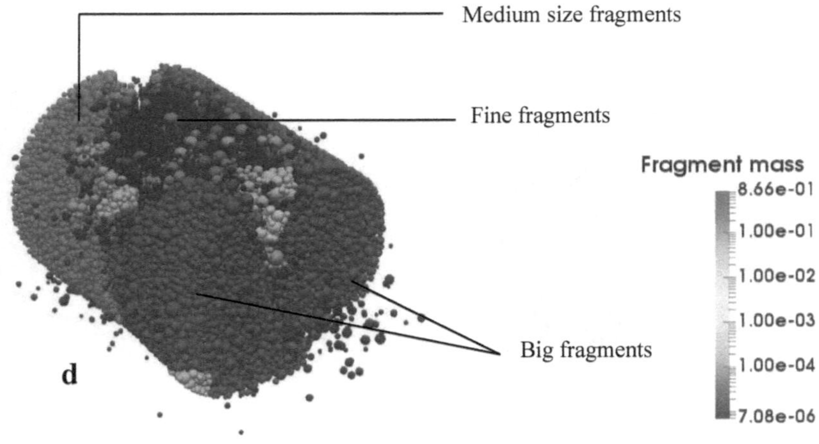

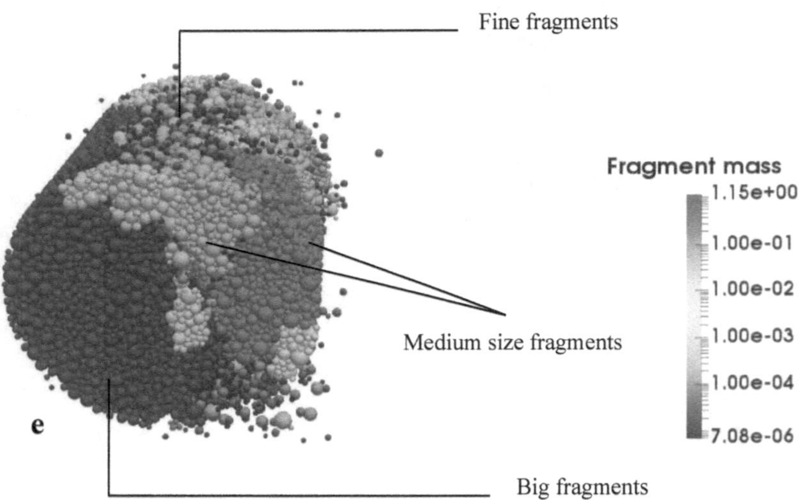

Fig. 7.11: Figures showing the ranges of fragments generated at the different crack levels. (a) fine and big fragments at 0 crack. (b) fine and big fragments at 30 cracks. (c) fine, medium size and big fragments at 90 cracks. (d) fine, medium size and big fragments at 180 cracks. (e) fine, medium size and big fragments at 300 cracks.

7.2   Investigation of mineralogical composition

The current BPM-DEM approach was also adopted to construct numerical rock specimens in which the mineralogy was examined. As an initial effort, a binary sample was investigated i.e., two-phase minerals (valuable mineral and the gangue) for simplicity and to minimise possible variations that might be caused by complex ore texture. Furthermore, this methodology attempted to mimic a "simplified" UG2 mineral ore due to the notable two-phase mineral feature customarily associated with this ore. For UG2, Platinum group elements (PGEs) typically form weak and strong associations with other minerals (McLaren and De Villiers, 1982 and Schouwstra et al., 2000) which made it a suitable representation of a binary sample. Additionally, there was abundant breakage data available for UG2 including the study by Chikochi (2017). This work investigated ores such as the binary-phase UG2-spotted anorthosite shown in Fig. 7.12. This is composed mainly of about 66% Plagioclase and approximately 27% Orthopyroxene (Chikochi, 2017)

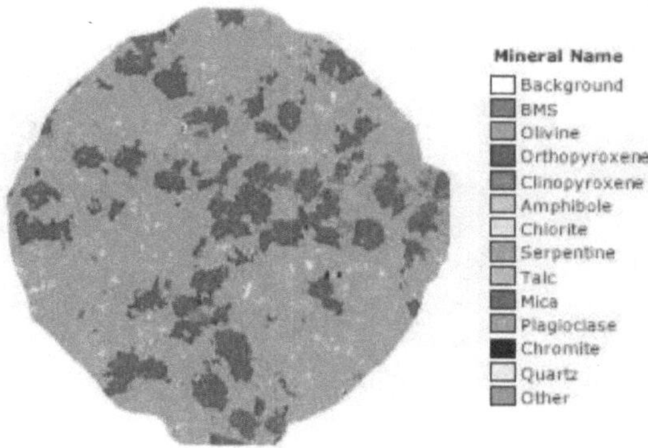

Fig. 7.12: QEMSCAN false image of UG2-spotted anorthosite. Extracted from Chikochi (2017)

7.2.1   Construction of DEM rock specimens with mineralogical composition

In this study, an attempt was made to mimic a "simplified" UG2 ore. A simplified UG2 ore sample is defined as a rock specimen with two-phase minerals in which one of the phases has weak interaction between grains and the other phase exhibits a strong interaction between grains. It was assumed that the soft component with weak interactions be the "valuable mineral" while the other phase is called the "gangue". The concept of building the numerical mineralogical rock specimen entails the following four steps.

- Firstly, generate a homogeneous rock specimen.

- Specify random DEM-spheres known as "seed points" within the homogeneous specimen to represent mineral grain points

- Group spheres closest to these seed points together.

- Specify or vary the percentage of the valuable or gangue material in the grouped mineral grains to produce different mineral grades

The above-highlighted steps taken to construct the simplified UG2 rock specimen with mineralogical compositions are illustrated in Fig. 7.13a-d. Fig. 7.13a represents the homogeneous specimen, while Fig. 7.13b shows the homogeneous specimens with seed points (coloured spheres). Closest spheres grouped to each seed points are shown in Fig. 7.13c and Fig. 7.13d shows the percentage of the minerals that were defined to be valuable (red-grouped minerals) and gangue (light-grey-grouped minerals). A 20% grade (red-grouped minerals) was used as an illustration in this case.

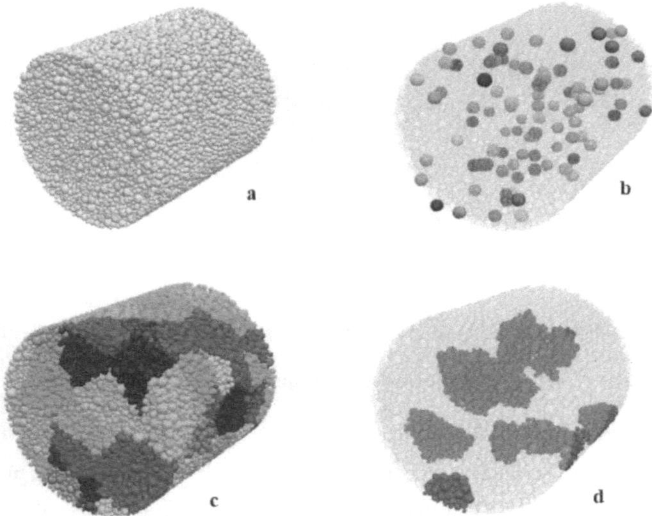

Fig. 7.13: An example of a cylindrical specimen with mineralogical compositions. (a) represents the homogeneous specimen, (b) shows the homogeneous specimen with seed points (coloured spheres), (c) spheres nearest to each seed point that were grouped together. (d) shows the percentage of the minerals that were defined to be valuable (red-grouped minerals) and gangue (light-grey-grouped minerals).

In summary, and as a proof of concept for this methodology to examine the effect of mineralogical composition, three main assumptions were made to construct numerical rock specimens which could be approximated to mimic a UG2-ore.

- Firstly, rock specimens are made up of a two-phase mineral.

- Secondly, one phase has mechanical properties that are twice the value of the other phase i.e., the strong component and the weak component. In this study, the weak components were taken as a valuable mineral (red-grouped minerals) while the strong component constitutes the gangue (light-grey-grouped minerals). Therefore, the percentage grade composition of the mineral corresponds to the direct proportion of the valuable minerals

- Lastly, the value of the mechanical properties of the interconnecting interaction between the strong and weak component has the same mechanical properties as the weak component.

7.2.2 Dependence of fracture behaviour on mineralogical composition

The effect of increasing the percentage composition of the grade was investigated at 6 levels which are; 0 (homogeneous), 10, 25, 50, 75 and 100%. Fig. 7.14 a-f highlight the typical distribution of valuable minerals within the rock specimens at different mineral grades. These distributions and locations of grade followed a random distribution per specimen. As such, 30 repeats of impact breakage simulations were conducted at each percentage of grade composition. This is to minimise the variation of fracture measures due to the random packing of DEM-spheres and also to reduce variation from the random location and distribution of the valuable minerals.

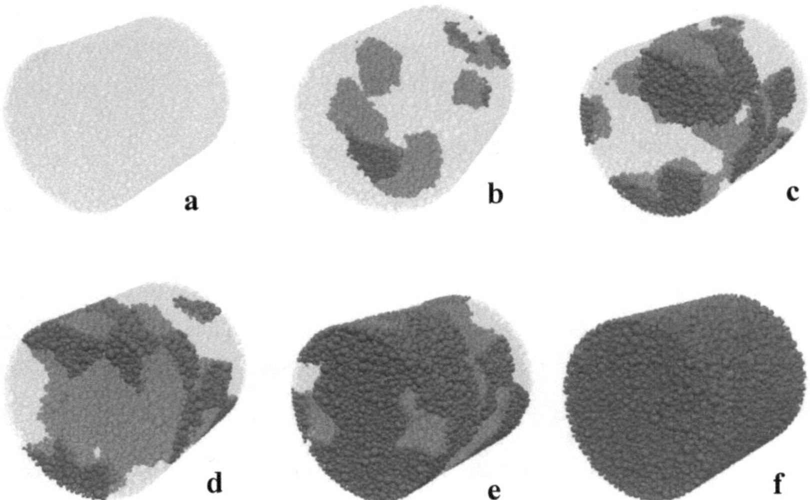

Fig. 7.14: Typical mineral grades of rock specimen. (a) 0% ,(b) 10% , (c) 25%, (d) 50% , (e) 75% and (f) 100% grade composition

Fig. 7.15 shows the variation of the measured fracture as the percentage of the grade increases. Homogeneous specimens with zero grade composition i.e., without any inherent weak-phase mineral have the highest fracture force of about 1.2kN. This decreases as the composition of the grade increases. Specimens which are entirely composed of the weaker grain (i.e. 100% grade) have the least fracture force of less than 0.1kN.

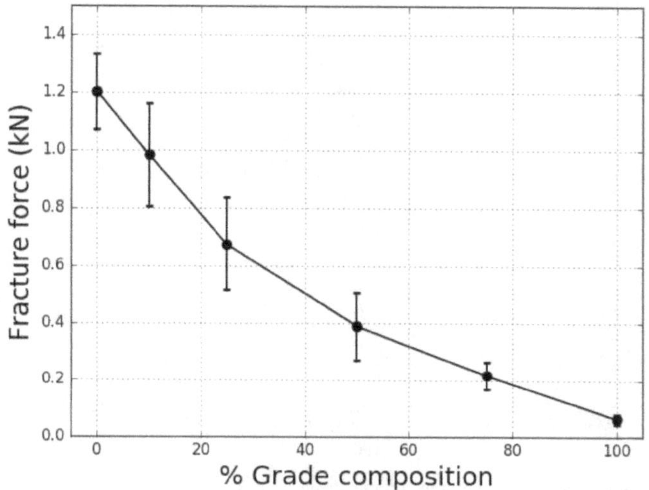

Fig. 7.15: Variation of percentage mineral grade against the measured fracture force.

7.2.3   Breakage behaviour

Further investigations were conducted on a representative specimen at each grade composition to understand the breakage behaviour and liberation of the valuable minerals. A typical force-time history for different grades is shown in Fig. 7.16. While the peak force decreases as the grade composition increases, it is further observed that there tends to be more oscillation of the force-wave in the presence of the valuable mineral. These oscillations are noticeable at even the least grade of 10%. This can be attributed to the introduction of a second phase through which the stress transmitted within the specimen becomes more asymmetrically distributed by having a weakened mineral phase along its path. Thus, the stresses are transmitted more randomly and less uniformly. This effect, in turn, becomes unnoticed for the 100% grade as this is returns to be a homogeneous form of the specimen.

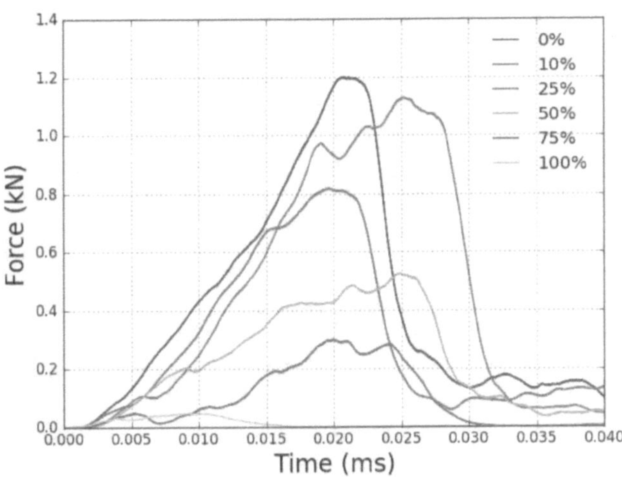

Fig. 7.16: Force-time history of the representative specimen at different percentage grade composition

The fracture paths were further examined to consider preferential exposure (liberation) of a phase from the other mineral phase. Fig. 7.17 (a-r) summarises the locations and distributions of the valuable mineral, the fracture paths and locations within the rock specimen as well as fragmentations obtained at different mineral compositions. At 0% grade composition (Fig. 7.17a), the fracture path travels almost vertically downward (Fig. 7.17b) with the specimen spitting into almost two halves (Fig. 7.17c). In the presence of 10% grade valuable minerals (Fig. 7.17d), fracture paths tend to flow around the valuable minerals (Fig. 7.17e) which results in an irregular fragmentation (Fig. 7.17f). Furthermore, at 25% and 50% grade composition (Fig. 7.17g and j), the fracture paths and directions seem to be controlled and directed along the densely populated regions of valuable minerals which give rise to the tributaries of the fracture patterns (Fig. 7.17h and k) and finally results in irregular fragmentation of specimens (Fig. 7.17i and l). However, at 75% and 100% grade composition (Fig. 7.17m and p), when the rock specimen is predominantly composed of valuable minerals, there is no defined path of the fracture. Instead, the fracture locations are randomly distributed (Fig. 7.17n and q). This anomalous behaviour is attributed to the overall deviation of the mechanical property of the rock specimen from brittle-elastic to a more elastic nature. The consequence of this is a compression and vibration of the rock specimen in the impact zone which led to splashing of some of the DEM-spheres at these fracture locations (Fig. 7.17o and r).

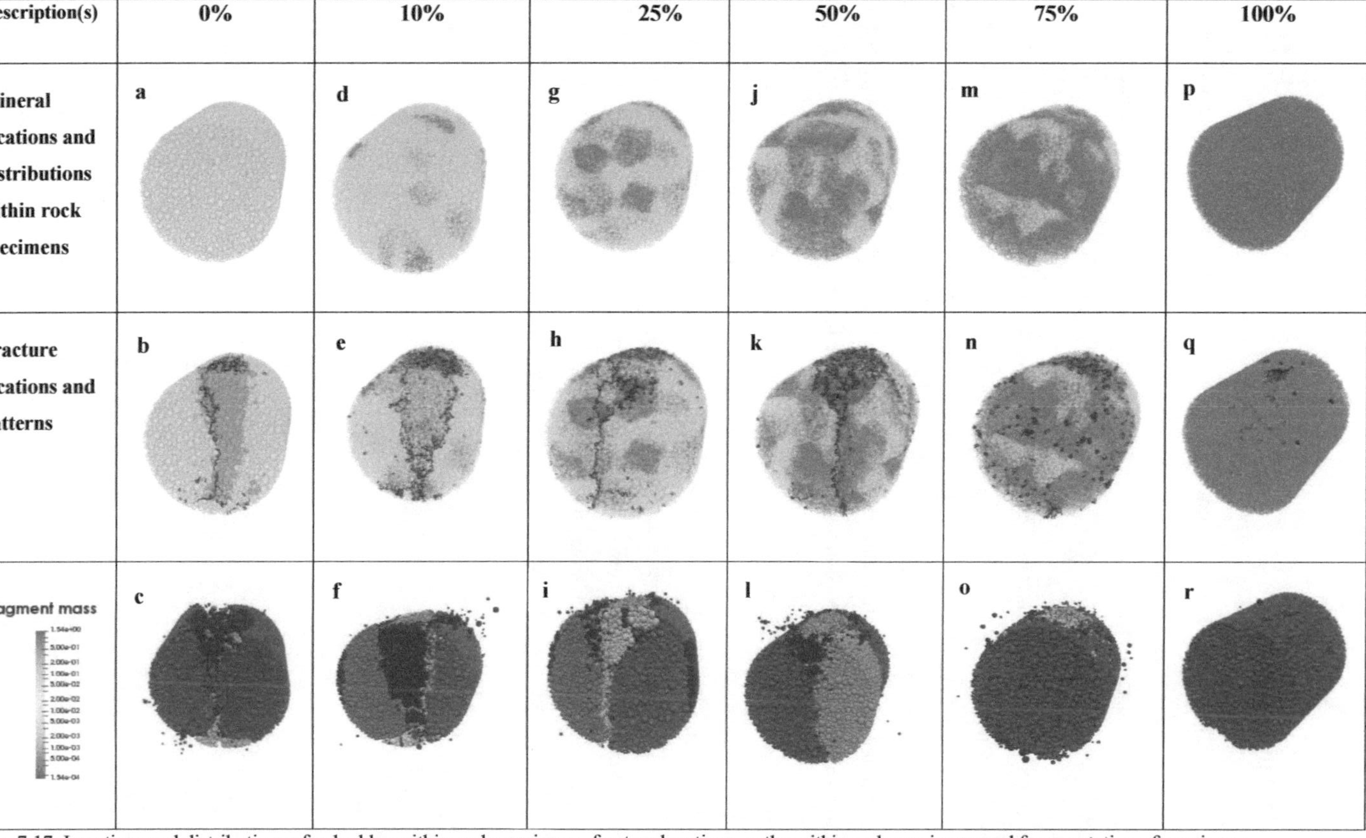

Description(s)	0%	10%	25%	50%	75%	100%
Mineral locations and distributions within rock specimens	a	d	g	j	m	p
Fracture locations and patterns	b	e	h	k	n	q
Fragment mass	c	f	i	l	o	r

Fig. 7.17: Locations and distributions of valuables within rock specimens, fracture locations, paths within rock specimens and fragmentation of specimens

7.3   Chapter summary

The current BPM-DEM methodology was used to investigate two important factors that are impractical to study by experiments due to the inherent variabilities typically encountered during rock breakage tests. These are the effect of pre-weakening within rock specimens and mineralogical composition.

The current chapter demonstrated how the levels of pre-damage in specimens could be numerically quantified in terms of the number of pre-existing cracks, fracture surface area as well as the new fracture surface area generated after impact. An increasing number of pre-existing cracks in rock specimens yielded a lower fracture force and ultimately produced a lower amount of new fracture surface area. Consequently, the irregularities of the fracture patterns and fragmentation become more apparent as the pre-damage increases.

For the investigation on the effect of mineralogical composition, a binary phase rock specimen was considered, i.e., strong and weak phase. It was assumed that the strong phase is the gangue mineral and was prescribed with twice the mechanical properties of the valuable minerals. The proportion of the valuable minerals corresponds to the percentage grade of the mineral. Simulation results showed that there is a decrease in the measured fracture force increases as the percentage composition of the grade increases. Furthermore, fracture paths tend to be favoured along the regions of valuable minerals.

8 General discussion and Novelty

The discussions on the results that were presented in previous chapters are as follows:

- Calibration of the BPM parameters against macroscopic mechanical properties. Results were presented in chapter 3 and the discussion is presented in section 8.1.

- Evaluation of the robustness of BPM in DEM for simulating rock breakage in a SILC device in order to determine its suitability for the current study. These results were presented in chapter 5 and a follow-up discussion on these are presented in section 8.2.

- A comparative study of impact breakage of single rock specimen using the BPM-DEM methodology and SILC experiments conducted by Barbosa et al. (2019). The result of this was presented in chapter 6 and the discussion is presented in section 8.3

- The results on simulated breakage behaviour of a single rock specimen were presented in section 4.3 while the results on the independent investigations of the effect of pre-existing cracks and mineralogical compositions on load response were shown chapter 7. Discussions following these are reported in section 8.4.

8.1 Calibration of the bonded particle model

The systematic parametric sensitivity study (SPSS) was used to investigate the forms of relationships between model parameters and macroscopic mechanical properties. Similar to the DoE approach which develops a statistical model between parameters, an empirical calibration relationship was also developed. However, the SPSS was found to be more robust which deemed it to be more advantageous for this work.

Firstly, the interval or variation of the model parameters can be closely controlled over a wide range. This ensures the minimisation of error in terms of over/underestimation in the developed model prediction. This differs from the DoE approach, which commonly entails selecting a high, medium and low levels of variables. Other points within these levels are usually statistically predicted points. Secondly, the simulated data points and the developed calibration relationships both exhibited the linear and quadratic effect of description also similar to the DoE approach. This can be seen with a flat surface in the relationship between

Young's modulus, elastic beam modulus and Poisson's ratio under compression (Fig. 3.12) and tension (Fig. 3.13). A curvilinear surface plot exists between the unconfined compressive strength, cohesive strength and internal angle of friction (Fig. 3.14) as well as the unconfined tensile strength, cohesive strength and internal angle of friction (Fig. 3.15). These points give emphasise on the robustness and the validity of this approach to capture these effects which are in line with the commonly known approaches.

It is also important to stress that the spatial distribution of DEM-spheres selected for the current calibration study worked consistently when examining fundamental breakage studies (quasi-static loading) as well as dynamic breakage scenarios (SILC experiments). It should be noted however that a different variation in spatial distribution may result in slight differences in calculated macroscopic mechanical properties.

8.2 Evaluation of the bonded particle model

In most prior studies in which breakage simulations have been conducted using a similar methodology, the sensitivity of the model resolution has not been given much attention. The number of discrete entities (DEM-spheres) does not realistically depict what could be contained in a typical rock but is nevertheless a key consideration when using BPM. The current investigation highlights that a high model resolution has a significant effect on the reduction of deviations and errors associated with the measured fracture characteristics, indicating that it is worth considering whether a selected resolution is numerically "safe" to operate at, to obtain realistic and consistent results.

A recommended approach to refine this methodology would be to integrate numerical results with mineralogical studies such as X-ray computed tomography data (Ghobani et al., 2011), whereby an approximated distribution of grain sizes could be used as a basis to define the *Rmin* and *Rmax* parameters as well as the distribution used in simulations. The current work assumed a constant ratio of *Rmin* and *Rmax* for simplicity.

Size dependency is another criterion to check the sensitivity of the model. The rock specimen stiffness showed a dependency on the size of the specimen, increasing with specimen size. It is generally expected that the reverse would be observed, as larger specimens increase the likelihood of containing flaws and other weaknesses. In this investigation, rock specimens were modelled as entirely homogeneous, such that based on mathematical considerations of linear elastic theory (section 4), stiffness increases by a factor of the ratio of the area to specimen length multiplied by Young's modulus (Eq. 4.2).

The variation in fracture characteristics with Young's modulus and the UCS of the rock specimen describe the elastic-brittle nature of a wide range of rock types. This is in agreement with linear elastic theory and the cohesive failure criterion used in this model. The Young's modulus and UCS were considered to be the two most important parameters for this application of applied load on the rock. However, the approach demonstrated with the simulation sweep over the applicable range of values can be extended to other scenarios to quantify the material response against known macro strength measurements.

8.3 Comparative study

The BPM of rock in DEM has shown to have a close resemblance to homogeneous synthetic rock specimens. Synthetic rock specimens were composed of silica sand and binders holding the silica sands together (Barbosa et al., 2019). This can be likened to the numerical rock specimen where the silica sands represent the DEM-spheres and binders represent the brittle-elastic beam of the BPM rock. This made it be considered an ideal specimen for comparison.

The key consideration is the consistency of DEM results with experiments. Here, DEM consistently predicted the experimental fracture force and fracture patterns for both cylindrical and spherical rock specimens. It was, however, noted that there was a slightly reduced value of about 0.2kN for the predicted force relative to the measured fracture force for the sphere in comparison to the cylinder. This is on an opinion attributed to possible variation in the mechanical properties derived during the printing of this shape. On a second opinion, it is not impossible that the estimated cohesive strength from the UCS calibration relationship accounts for this slight deviation. In order to address and test the validity of this conjecture numerically, the cohesive strength was tuned at the experimental Young's modulus by following the steps based on a trial and error approach:

- Select a random cylindrical specimen from those used in the experiment as a reference sample size. In this case, a 7.3mm cylinder was selected.

- Conduct SILC simulations on the selected sample at the measured Young's modulus and, while varying the cohesive strength (10MPa -70MPa), measure the fracture force.

- Compare the measured fracture force for each cohesive strength from DEM simulations against experiments for either vertically printed or horizontally printed cylinders. The value of the cohesive strength where the fracture force of the simulation matches with experiment defines the calibration point for the rock specimen.

- Using the calibration points, conduct a size dependency test and compare against experiments.

Table 8.1 summarises the result of the calibration exercise showing the measured fracture force for both experiment and DEM simulation using the 7.3mm cylinder. Fig. 8.1 presents the results of the cohesive strength calibration for the cylinder. The calibration point (cohesive strength value) for which the measured fracture force from the DEM simulation matches the experiment is 40MPa.

Table 8.1: A summarised result of the calibration exercise

Reference size (mm)	Exp. Fracture force (kN)	DEM Fracture force (kN)	Calibrated cohesive strength (MPa)
7.3	0.519	0.496	40

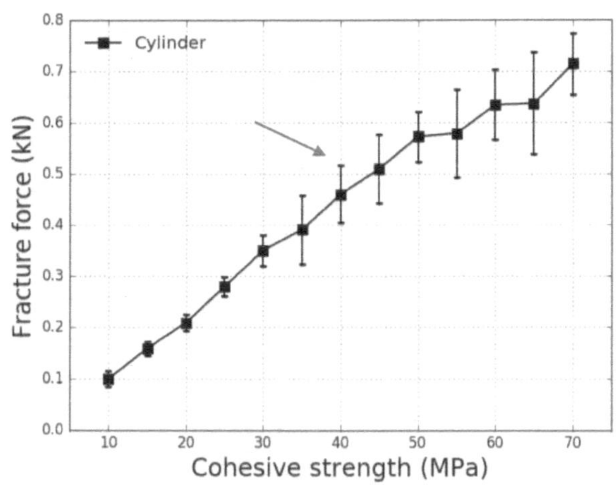

Fig. 8.1: Fracture force vs cohesive strength for the reference cylinder (vertically printed cylinder).

Fig. 8.2 presents the result of the size dependency test for the vertically, horizontally printed cylinders and spherical specimens using the 40MPa cohesive strength value. This "manual" tuning of the cohesive strength shows that there is a closer prediction of the fracture force by DEM simulations.

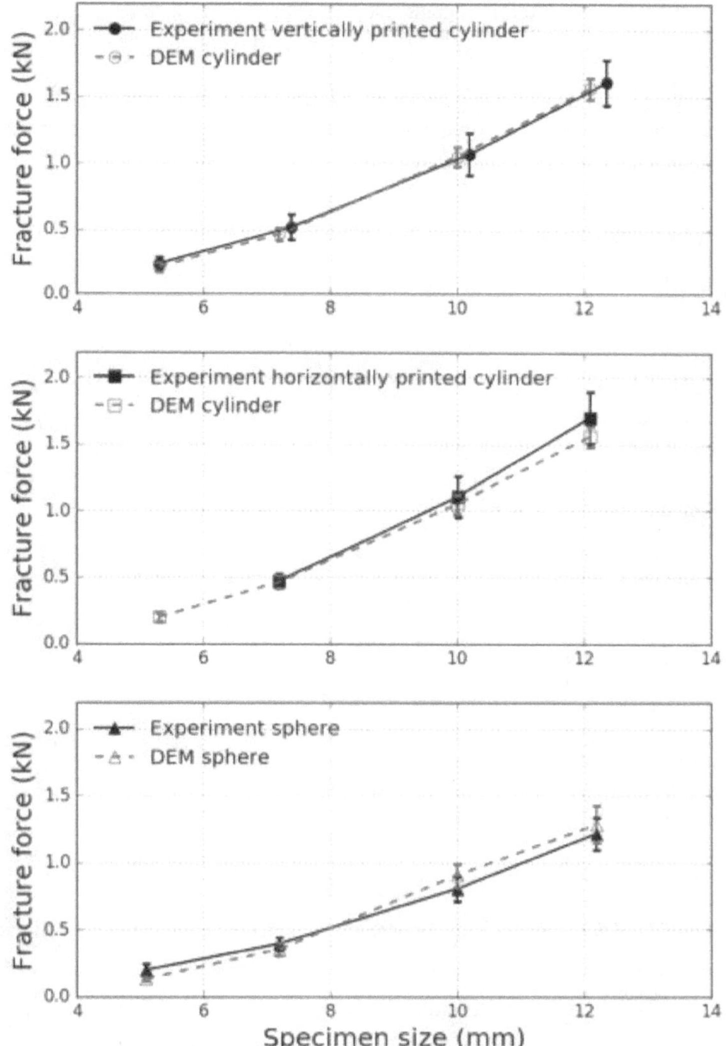

Fig. 8.2 Fracture force measured for different sizes for the vertically printed cylinder (a), horizontally printed cylinder (b) and sphere (c) with the same cohesive strength (40MPa)

It also follows that the mechanical properties of a homogeneous rock specimen should be independent of shape and size, forming another basis to validate the current BPM-DEM methodology. Consequently, the calibrated mechanical properties would not need to be altered to suit a particular shape or size since the material properties should remain the same.

Finally, a plot of the SILC fracture force against the DEM simulation fracture force for all the sizes of the synthetic specimens is shown in Fig. 8.3. There is a direct correlation between both cases with the y=x. The R-square value of 0.9923 quantifies the strong agreement between the experiments and the simulations which suggest that the model predicts reasonably well.

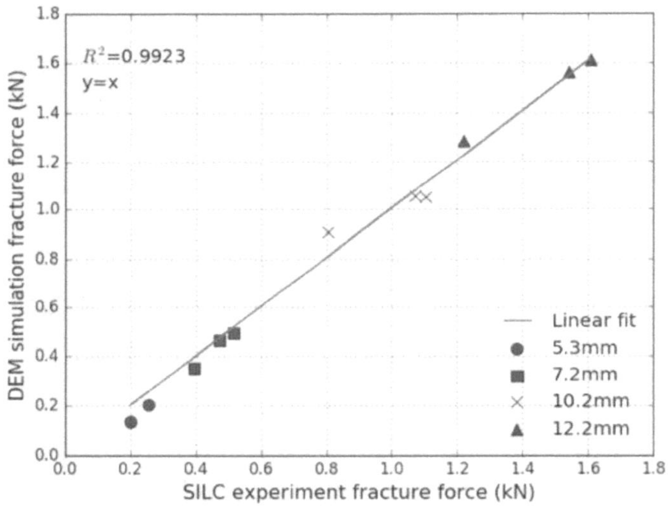

Fig. 8.3: SILC experiment fracture force vs DEM simulation fracture force

8.4   Single rock specimen breakage study

8.4.1   Discussion on simulated breakage behaviour

The examination of the DEM breakage of a single rock specimen showed typical behaviour characteristic of a SILC experiment. The force vs. time obtained from the simulation could be used to extract standard measures of the material response under load. The advantages of numerical simulations were in being able to further interrogate the micro-scale behaviour within the rock specimen during the loading event. A key observation, for instance, was that a low percentage of approximately 9% of connecting bonds were broken when the rock specimen yielded under load. This was confirmed by visualization of the fragmentation during the impact event which indicated that large segments of the specimen remained intact. Such results can be used to emphasize and study the attributable causes of inefficiency in comminution practices, whereby significant amounts of energy can be expended to achieve minimal gains in respect of enhancing liberation. It is worth highlighting too that the key findings of this work agree with previous work on impact breakage (Bonfils, 2017; Jiménez-Herrera et al., 2017; Williams, 2014; Tavares, 2004; Thornton et al., 1999).

It is also worth noting, that even though the SILC test occurs under a dynamic loading compared to the UCS test, which is quasi-static, key features customarily associated with elastic-brittle strain were observed during the initial loading period (zone A). Finally, the stress distribution within the rock specimen was determined. The high-stress intensity zones at the contact points (impactor-rock specimen and rock specimen-wall) initiated fracture locations which under increasing stress led to eventual failure.

8.4.2   Discussion on pre-weakening

Five levels of pre-damage quantified by the amount of pre-existing cracks/ fracture surface area in the homogeneous rock specimens were created (Table 7.1). These numerical pre-weakened rock specimens attempted to replicate rock specimens that could have gone through some weakening prior to the comminution stage. Results of the fracture force and the new fractured surface (Fig. 7.4) showed a linear trend when plotted against the level of damage. It has been previously shown that the fracture force is proportional to the fracture energy (Section 4.2.3). This result indicated that this current methodology could show a similar trend to such experiments.

The increasing number of major fragments generated at a high number of cracks (300 cracks) suggested that the pre-existing cracks played a more dominant role in crack propagation than the propagation from the point of impact. The fracture patterns followed conventional fracture mechanics principles. For homogeneous specimens, cracks initiated at points of the most intense stress and propagated to failure. Hence, rock specimens typically cleaved into two major fragments (Fig. 7.11 a and b). As the crack density increased, the tendency of rock specimen to split into two decreased (Fig. 7.11c-e). This indicated that the presence of damage greatly influenced the likelihood and nature of the fracture, as known from Griffith's theory (Section 2.1.2). More fragments were generated as the crack density increased, due to the increasing tendency for the specimen to splinter along these flaws. Brown and Jones (1996) found that rock specimens with increasing numbers of flaws yielded more readily to impact and generated more fragments; the current work follows this observation.

More closely related to this finding is the study conducted by Haeri et al. (2015). Authors conducted compressive load experiments and numerical studies on moulded synthetic rock specimens with 5 levels of major cracks namely; 1, 2, 3, 4 and 5 cracks as shown in Fig. 8.4. Although cracks were introduced in a well-defined (parallel) pattern rather than random arrangements typical of real rocks, results also confirmed that the measured breakage load of the specimen decreases as the number of cracks increases as shown in Fig. 8.5. However, the propagation of the fracture lines from the point of loading was intercepted by the presence of the major cracks. It should be recalled that the compressive load was applied from both the top and bottom of the specimen here (Haeri et al. 2015). As such, fracture lines tended to move slightly diagonally from the top load to the right and slightly diagonally from the bottom load to the left before an interception by the presence of cracks along their path as shown in Fig. 8.6 a -e.

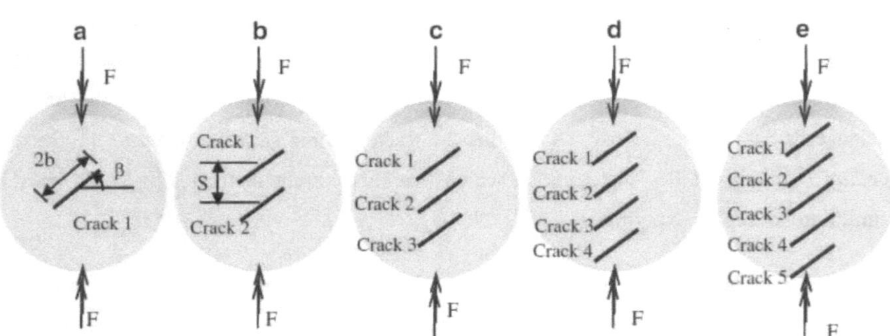

Fig. 8.4: Schematic representation of rock specimens showing the pre-existing crack arrangement for experimental and numerical study (Haeri et al. 2015).

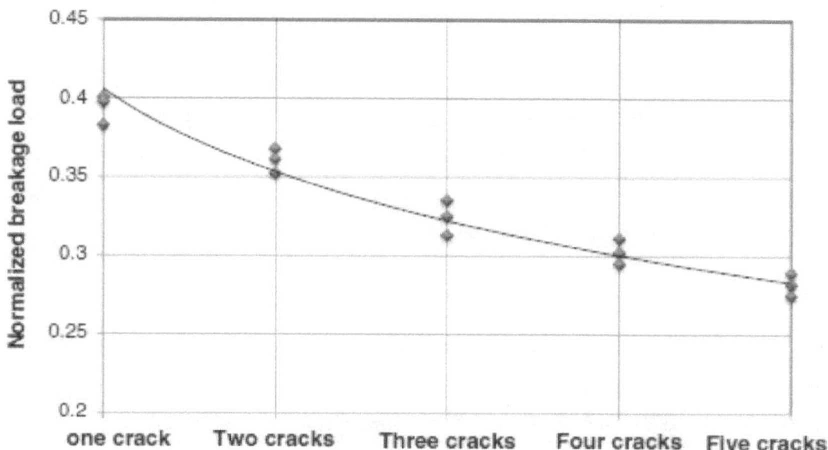

Fig. 8.5: breakage load response versus the number of cracks (Haeri et al., 2015).

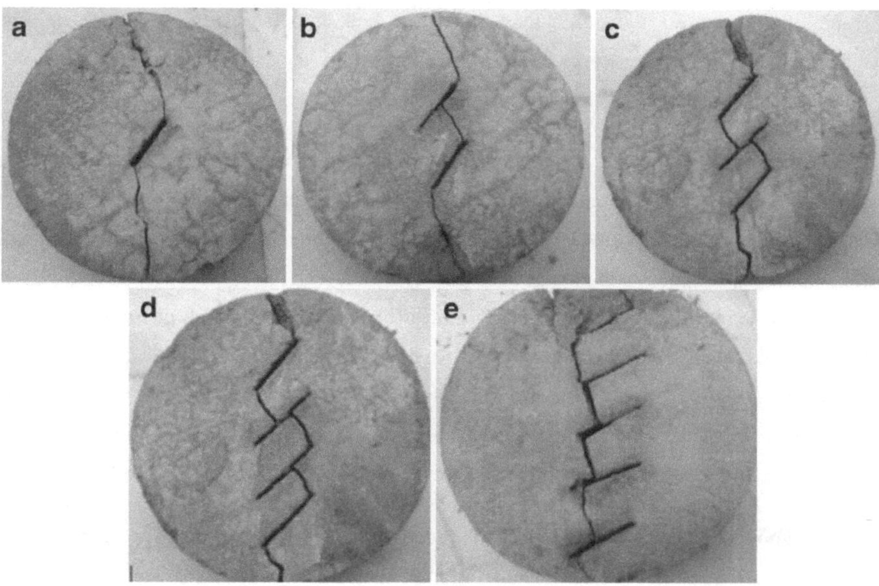

Fig. 8.6: Fracture path of specimen containing (a) 1 crack, (b) 2 parallel cracks (c) 3 parallel cracks (d) 4 parallel cracks and (e) 5 parallel cracks .

Several authors (Krogh, 1980; Pauw and Maré, 1988 and Bonfils et al., 2016) have reported that such low-stress level impacts cause weakening and formation of cracks. Results herein suggest that the level of weakening can be estimated and quantified with the new fracture surface area generated in relation to the input or fracture energy. With this information, one can compare the energy expended in causing breakage by single impacts to that over several impacts, basing this with respect to its efficiency in causing the desired level of fragmentation. Such results also emphasize some of the attributable causes of variation observed during ore hardness characterization tests and in situ whereby the pre-weakening affects the breakage behaviour observed.

8.4.3 Discussion on mineralogical composition

It is known that ore texture provides vital information on ore behaviour to assist in better circuit operation, design and planning (Rule and Schouwstra, 2011). Largely, the specimen preparation procedure involved in the analysing the texture of an ore specimen (e.g. by QEMSCAN) involves abrading and polishing the surface. Additionally, the analysis is conducted only on the exposed surface i.e., 2-dimensions (2D) to the equipment, with the result of the analysis then used to estimate the bulk mineralogy in the specimen. The numerical approach taken to examine mineralogy offers a means of supplementing such information by a quantifiable study of the entire specimen.

It is worth highlighting the assumptions that were made to simplify the methodology to initially demonstrate its potential application. A binary ore phase was used as a simple test case and to mimic a simplified UG2 -ore. The fracture of the rock specimen along the path of the weaker phase suggests that crack propagation is enhanced under the presence of a weaker phase. This leads to a more exposed surface area (liberation) of the weak phase. However, at higher grade when the composition of the weak component is significantly high, the rock specimen tended to behave like an elastic material. This was largely a limitation of the parameters selected to represent the weaker phase, in later work the possible range for mechanical properties of a typical rock can be ascertained.

It is acknowledged that more work is needed in this regard to improve some of the assumptions that have been made. For instance, the ratio of the mechanical properties of valuable material to gangue is characteristic of real UG2-spotted anorthosite and no experimental data was available. Furthermore, the mechanical properties at the interface between the gangue and valuable were assumed to have the mechanical properties of the valuable. However, it should be emphasised that this methodology has demonstrated a framework towards using a numerical tool such as DEM to interrogate the effect of mineralogical composition during breakage.

8.5 Application to comminution

Following from the fundamental understanding of the key influencing factors of ore breakage, and variabilities from these factors, the current work has provided a promising means of examining the breakage process in the comminution environment. For example, consider a generic comminution circuit shown in Fig. 8.7. The power utilisation of either the SAG or ball mill is typically based on the quantity of ore loaded and the rotational speed. The utilization of this energy to achieve breakage is in turn approximated by empirical correlation against the target size achieved.

This approach shows little or no fundamental consideration for the mechanical properties of the ore, its level of pre-weakening or its mineralogy. It is known that preceding knowledge on the state and strength of the ore feed makes a significant contribution in determining the optimal approach to take for comminution and liberation. One such analytical technique is the Atomic Force Microscopy (AFM) which has the capability to benchmark output measures with numerical simulations. The AFM has the main advantage of measuring the mechanical properties relative to composition within a sample in addition to imaging the composition and manipulating the mechanical properties (Dejous et al., 2017). Coupled with the calibration methodology demonstrated in this work, a later study could investigate the potential of using the BPM-DEM to investigate the factors that affect the breakage behaviour of a typical comminution device.

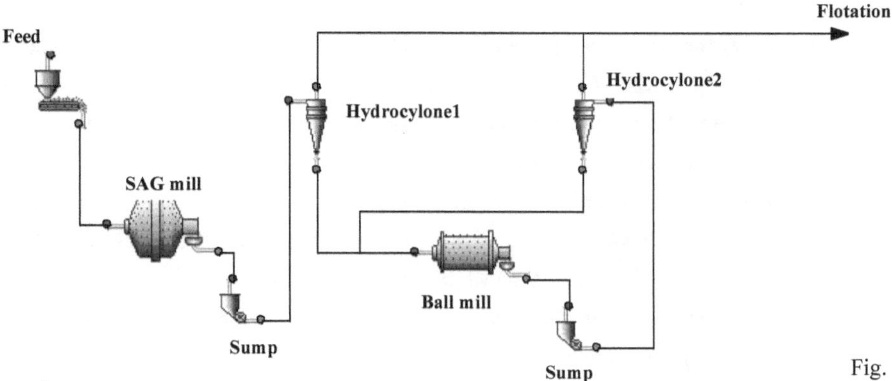

8.7: A typical comminution circuit (Mwale et al., 2018)

8.6 Novelty

The following areas of novelty have arisen from the current work:

- **A robust BPM-DEM approach.**
  A robust BPM-DEM approach using an extensive calibration methodology and mathematical relationships were developed. Chapter 3 described the DEM approach used to measure rock mechanical properties according to the ISRM standard as well as the developed calibration relationships proposed as a basis for estimating model parameters without conducting further calibration. An evaluation of the BPM demonstrated in chapter 5 emphasised the significance of model resolution variation on the standard deviations of fracture measures. Highest model resolution (highest number of DEM-spheres) with a close resemblance to real rock showed the least deviation on fracture measure. In addition, the number of repeats (30) per impact simulation significantly reduced the contributing effect the random packing of DEM-spheres for each rock specimen. Such a task from an experiment could be very challenging and time-consuming in addition to possible human errors.

- **Simulating breakage of synthetic rock specimen in mineral processing context.**
  The evaluated numerical SILC model was applied to simulate the breakage of synthetic rock specimens with material properties selected based upon the highlighted calibration methodology. The simulations predicted the measured fracture force and fracture behaviour reasonably well for distinct rock shapes and sizes without altering the mechanical parameters of the material.

- **Isolating the effect of pre-weakening and mineralogy on rock breakage.**
  Studying the effect of pre-existing cracks and mineralogy composition on load response was demonstrated in chapter 7. Isolating and quantifying pre-damage by experiment is impractical. The current approach offered a potential means of examining this important characteristic in closer detail. The fracture force decreased, and the extent of breakage increased as the level of pre-existing cracks was increased. Not only does this BPM-DEM model predicts rock breakage satisfactorily, it captures the variations of progeny size distributions in the presence of cracks. Additionally, a unique methodology was developed and tested to simulate the breakage of a binary

phase mineral to investigate the effect of ore mineralogy. This approach can be extended to a more complex phase mineral.

9 Conclusions and Recommendations

This work has taken a step to extensively use the bonded particle model (BPM) in DEM to study the mechanism of breakage of a single rock specimen in a short impact load cell device (SILC). Furthermore, the effect of pre-weakening and mineralogy are independently tested. The following section reports the conclusions and recommendations for future work.

9.1 Conclusions

9.1.1 Calibration of the BPM

The BPM-DEM was demonstrated as a promising technique, capable of predicting the results of controlled laboratory experiments under similar operating conditions. It was shown that the BPM parameters can calibrate macroscopic mechanical properties and perform breakage simulations in line with standard geotechnical tests. It was further shown that empirical calibration relations can be obtained to relate BPM model parameters to macroscopic mechanical properties. These calibration relations are then utilised to tune the BPM such that its macroscopic properties match those measured for synthetic rock specimens.

9.1.2 Evaluation of the BPM

It was demonstrated that the current BPM-DEM can generate data reasonably identical to experiment for the impact breakage of a rock specimen in a SILC device, having been accurately calibrated. This includes calculations of impact force, free surface, stress data as well as the qualitative fracture patterns that are obtained from experiments. It is acknowledged that the calibration of the model is crucial as the initial step before carrying out simulations. The model resolution used for the BPM-DEM was also found to be an important consideration to reduce the variability of simulations. An increase in model resolution resulted in a decrease in the deviation obtained when calculating the key fracture measures.

A size dependency test with the selected model resolution showed an increase in fracture force as the size of the rock specimen increased. This is in agreement with the fundamentals of linear elastic theory, for which homogeneous specimens would be expected to increase in strength with size.

Both the macroscopic Young's modulus and unconfined compressive strength showed a consistent trend against the measured force which allowed for a calibration relationship to fit these mechanical properties such that desired specimen characteristics could be tuned. The approach shown in this work covered a wide range of typical rock types.

9.1.3   Comparative study

The application of the current methodology simulated impact breakage of cylindrical and spherical synthetic rock specimens and compared against experimental results conducted by Barbosa et al. (2019). The consistency in terms of the measured fracture force at different rock sizes and the number of fragments generated from the simulations matched quite well with the experimental results. This consistency further suggested that with a well-conducted calibration exercise and adequate selection of DEM model resolution, BPM-DEM can naturally predict the measured fracture force and fracture reasonably well for all rock shapes and sizes without altering the mechanical properties.

9.1.4   Single specimen breakage study

(a) Breakage behaviour

The capacity of the BPM-DEM approach to relating the micro-scale response (stress distribution, broken bonds, fracture locations) to the measured macroscopic responses (fracture force and fragmentations) was studied by considering a homogeneous rock specimen subjected to impact loading. The measured fracture force and fragmentation of rock specimen were informed by the stress distribution (stress tensor), percentage broken bonds and energy utilisation at segmented time zones. The measured macroscopic response was preceded by the interactions of these micro-scale responses. These findings also aligned with the hypothesis of Tavares (2007) which stated that *"rock breakage is influenced by the interaction of several complex micro-processes and properties such as stressing condition and comminution environment"*. This facilitates an improved understanding of the fundamentals of ore breakage.

(b) Pre-weakening

A micro-crack approach was employed to introduce pre-existing cracks into homogenenous rock specimens. The extent of pre-weakening was quantified in terms of either percentage number of broken bonds, pre-damage level or fracture surface area. As the pre-existing cracks increased, the force to fracture decreased as crack propagation is enhanced by the presence of pre-existing cracks. A linear relationship was established between the fracture force and new fracture surface area. This offers a potential means of assessing the benefit of pre-weakening in terms of the energy required to achieve a particular degree of breakage.

(c) Mineralogical composition

A simple binary-phase mineral was considered to represent the mineralogical composition of a rock specimen in an initial attempt to mimic the UG2-spotted anorthosite. The valuable mineral was assumed to be the soft component while the gangue constituted the hard component, specified with mechanical properties in a 1:2 ratio. The breakage results confirmed that the fracture force decreased as the amount of the weaker component increased. Furthermore, fracture paths tend to be directed towards the populated weak phase.

9.2   Recommendations

This work has employed a BPM-DEM approach to study impact breakage of both homogeneous and heterogeneous rock under impact loading in a SILC device. This has provided a solid framework with many avenues for further improvement from the outcomes. The following recommendations are made:

- To utilize process mineralogy data to inform and create specimens or more complex compositions. The breakage of such specimens can be carried out using the numerical methodology outlined in this work to examine characteristics such as the resistance to fracture and the likelihood of mineral exposure (liberation).

- To integrate the quantification of pre-existing cracks from X-ray CT scans into the numerical specimen and conduct breakage simulations in the manner presented. Many authors have attempted to describe realistic particle shapes for numerical studies using 3-D laser scans to mark geometrical points of rock specimen (Quist and Evertsson, 2016 and Jiménez et al., 2017). This was also attempted here as shown in Fig. 9.1 a-c. A rock specimen was scanned using a 3D laser scan (Fig. 9.1a), followed by a creation of a triangulated mesh points around the rock surface (Fig. 9.1b) and finally filling the triangulated surface with bonded DEM-spheres (Fig. 9.1c) similar to the description in section 3. These efforts in conjunction with further validation studies offer a promising avenue for future work.

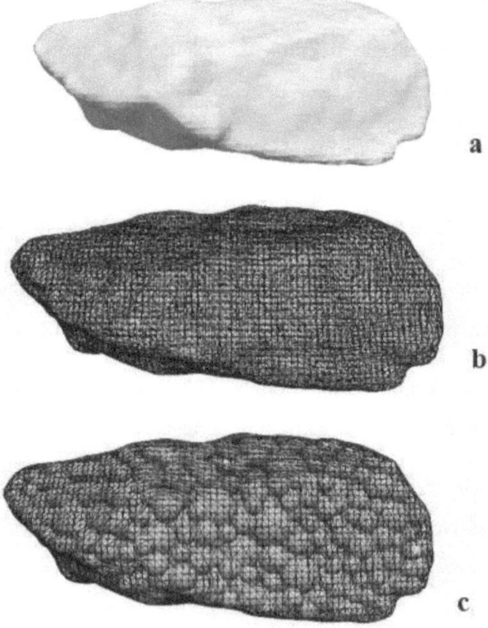

Fig. 9.1: (a) a 3D laser scanned rock (b) triangulated mesh points generated on the scanned rock (c) triangulated rock filled with DEM-spheres.

- To employ and integrate the use of an analytical technique such as the AFM to quantify the relative strength of mineral phases for breakage experiments.

- To employ simulation experiments using other breakage models and compare response such as the specific fracture energy with the current BPM-DEM model in order to validate the damage mechanics or incremental damage approach

References

Abe, S. & Mair, K. 2005. Grain fracture in 3D numerical simulations of granular shear. Geophysical Research Letters. 32(5).

Alibaba, 2019. 3D printer. Available: https://www.alibaba.com/product-detail/Best-Large-size-printing-house-building_62217360731.html?spm=a2700.drainage_lp_1.0.0.289a348bXyTasZ&s=p

Alnaggar, M. & Bhanot, N. 2018. A machine learning approach for the identification of the lattice discrete particle model parameters. Engineering Fracture Mechanics. 197:160-175.

American Society for Testing and Materials, 1991. ASTM D2938-86/95—Standard test method for unconfined compressive strength of intact rock core specimens

American Society for Testing and Materials, 2013. ASTM D7012-13—Standard Test Methods for Compressive Strength and Elastic Moduli of Intact Rock Core Specimens under Varying States of Stress and Temperatures. West Conshohocken, PA

Austin, L.G. 2002. A treatment of impact breakage of particles. Powder Technology. 126(1):85-90.

Ballantyne, G.R. & Powell, M.S. 2014. Benchmarking comminution energy consumption for the processing of copper and gold ores. Minerals Engineering. 65:109-114.

Barbosa, K., Chalaturnyk, R., Bonfils, B., Esterle, J. & Chen, Z. Testing Impact Load Cell Calculations of Material Fracture Toughness and Strength Using 3D-Printed Sandstone. Geotechnical and Geological Engineering.:1-32.

Baumgardt, S., Buss, B., May, P. & Schubert, H. 1975. On the comparison of results in single grain crushing under different kinds of load. Proc.11th Int.Miner. Process.Congr., Instituto Di Arte Mineraria E Preparazione Dei Minerali, Cagliari. :5-32.

Bbosa, L., Powell, M.S., Cloete, T.J. & Polke, R.F. 2006. An investigation of impact breakage of rocks using the split Hopkinson pressure bar. Journal of the South African Institute of Mining and Metallurgy. 106(4):291-296.

Bbosa, L.S. 2007. No title. Measurement of Impact Breakage Properties of Ore Particles using a Series of Devices.

Becker, M., Wightman, E.M. & Evans, C.L. 2016. Process mineralogy.

Bieniawski, Z.T. & Hawkes, I. 1978. Suggested methods for determining tensile strength of rock materials. International Journal of Rock Mechanics and Mining Sciences. 15: 99-103.

Bonfils, B. 2017. Quantifying of impact breakage of cylindrical rock particles on an impact load cell. International Journal of Mineral Processing. 161:1-6.

Bonfils, B., Ballantyne, G.R. & Powell, M.S. 2016. Developments in incremental rock breakage testing methodologies and modelling. International Journal of Mineral Processing. 152:16-25.

Bourgeois, F.S. & Banini, G.A. 2002. A portable load cell for in-situ ore impact breakage testing. International Journal of Mineral Processing. 65(1):31-54.

Brace, W.F. 1961. Dependence of fracture strength of rocks on grain size. The 4th US Symposium on Rock Mechanics (USRMS). American Rock Mechanics Association.

Briggs, C.A. & Bearman, R.A. 1996. An investigation of rock breakage and damage in comminution equipment. Minerals Engineering. 9(5):489-497.

Chaboche, J. 1988. Continuum damage mechanics: Part II—Damage growth, crack initiation, and crack growth. Journal of Applied Mechanics. 55(1):65-72.

Chandramohan, R., Holtham, P. & Powell, M. 2010. The influence of particle shape in rock fracture. XXV International Mineral Processing Congress. 6-10.

Charikinya, E. 2015. No title. Characterising the Effect of Microwave Treatment on Bioleaching of Course, Massive Sulphide Ore Particles.

Chehreghani, S., Noaparast, M., Rezai, B. & Shafaei, S.Z. 2017. Bonded-particle model calibration using response surface methodology. Particuology.

Chikochi, C. 2017. Ore Breakage Characterisation of UG2 Deposits using the JKRBT. Master's book. University of Cape Town

Cleary, P. 2001. Modelling comminution devices using DEM. International Journal for Numerical and Analytical Methods in Geomechanics. 25(1):83-105.

Cleary, P.W. 2000. DEM simulation of industrial particle flows: case studies of dragline excavators, mixing in tumblers and centrifugal mills. Powder Technology. 109(1-3):83-104.

Cropp, A. & Goodall, W. 2013. The influence of rock texture on mineral processing. Min Assist. :1-9.

Cundall, P.A. & Strack, O.D. 1979. A discrete numerical model for granular assemblies. Geotechnique. 29(1):47-65.

Dejous, C., Hallil, H., Raimbault, V., Rukkumani, R. & Yakhmi, J.V. 2017. Chapter 20 - Using microsensors to promote the development of innovative therapeutic nanostructures. Available: http://www.sciencedirect.com/science/article/pii/B9780323461429000207.

Delaney, G.W., Morrison, R.D., Sinnott, M.D., Cummins, S. & Cleary, P.W. 2015. DEM modelling of non-spherical particle breakage and flow in an industrial scale cone crusher. Minerals Engineering. 74:112-122.

Dube, T.T. 2017. Measuring the Fracture Energy of Bed Breakage using a Short Impact Load Cell. Master's book, Centre for Minerals Research, University of Cape Town.

DEM-Solution. 2011. EDEM 2.4 Theory Reference guide. Edinburg

Donze, F.V., Bouchez, J. & Magnier, S.A. 1997. Modelling fractures in rock blasting. International Journal of Rock Mechanics and Mining Sciences. 34(8):1153-1163.

Eberhardt, E., Stead, D. & Stimpson, B. 1999. Effects of specimen disturbance on the stress-induced microfracturing characteristics of brittle rock. Canadian Geotechnical Journal. 36(2):239-250.

Eberhardt, E., 1998, Brittle Rock Fracture and Progressive Damage in Uniaxial Compression, PhD book, Department of Geological Sciences, University of Saakatchewan.

Erdogan, F. 2000. Fracture mechanics. International journal of solids and structures. 37(1-2):171-183.

Eksi, D., Benzer, A.H., Sargin, A. & Genc, O. 2011. A new method for determination of fine particle breakage. Minerals Engineering. 24(3):216-220.

Fakhimi, A. & Villegas, T. 2007. Application of dimensional analysis in calibration of a discrete element model for rock deformation and fracture. Rock Mechanics and Rock Engineering. 40(2):193.

Fandrich, R.G., Clout, J. & Bourgeois, F.S. 1998. The CSIRO Hopkinson bar facility for large diameter particle breakage. Minerals Engineering. 11(9):861-869.

Freund, L.B. 1972. Crack propagation in an elastic solid subjected to general loading—I. Constant rate of extension. Journal of the Mechanics and Physics of Solids. 20(3):129-140

Fuerstenau, M.C. 2003. Principles of mineral processing. SME.

Gell, E.M., Walley, S.M. & Braithwaite, C.H. 2019. Review of the Validity of the Use of Artificial Specimens for Characterizing the Mechanical Properties of Rocks. Rock Mechanics and Rock Engineering. :1-13.

Genc, O., Ergon, L. & Benzer, H. 2004. Single particle impact breakage characterisation of materials by drop weight testing. Fizykochemiczne Problemy Mineralurgii/Physicochemical Problems of Mineral Processing. (38):241-255.

Ghorbani, Y., Becker, M., Petersen, J., Morar, S.H., Mainza, A. & Franzidis, J. 2011. Use of X-ray computed tomography to investigate crack distribution and mineral dissemination in sphalerite ore particles. Minerals Engineering. 24(12):1249-1257.

Goodman, R.E. 1989. Introduction to rock mechanics. Wiley New York.

Gray, G.T. 2000. Classic Split-Hopkinson Pressure Bar Testing. Materials Park, OH: ASM International, 2000. :462-476.

Griffith, A.A. 1920. The phenomena of flow and rupture in solids: Phil. Trans.Roy.Soc.Lond.Ser.A. 221:163-198.

Griffith, A.A. 1921. VI. The phenomena of rupture and flow in solids. Philosophical Transactions of the Royal Society of London.Series A, Containing Papers of a Mathematical or Physical Character. 221(582-593):163-198.

Griffiths, D.V. & Mustoe, G.G. 2001. Modelling of elastic continua using a grillage of structural elements based on discrete element concepts. International Journal for Numerical Methods in Engineering. 50(7):1759-1775.

Haeri, H., Khaloo, A. & Marji, M.F. 2015. Experimental and numerical analysis of Brazilian discs with multiple parallel cracks. Arabian Journal of Geosciences. 8(8):5897-5908.

Hahne, R., Pålsson, B.I. & Samskog, P. 2003. Ore characterisation for—and simulation of—primary autogenous grinding. Minerals Engineering. 16(1):13-19.

Han, Z., Weatherley, D. & Puscasu, R. 2015. Application of Discrete Element Method to Model Crack Propagation. 13th ISRM International Congress of Rock Mechanics. International Society for Rock Mechanics.

Han, Z., Weatherley, D. & Puscasu, R. 2017. A relationship between tensile strength and loading stress governing the onset of mode I crack propagation obtained via numerical investigations using a bonded particle model. International Journal for Numerical and Analytical Methods in Geomechanics. 41(18):1979-1991.

Hentz, S., Donzé, F.V. & Daudeville, L. 2004. Discrete element modelling of concrete submitted to dynamic loading at high strain rates. Computers & Structures. 82(29-30):2509-2524.

Herbst, J.A., Lo, Y.C. & Flintoff, B. 2003. Size reduction and liberation. Principles of Mineral Processing.SME, Littleton, CO. :61-118.

Herbst, J.A. & Potapov, A.V. 2004. Making a discrete grain breakage model practical for comminution equipment performance simulation. Powder Technology. 143:144-150.

Hodder K. 2017. Fabrication, characterization and performance of 3D-printed sandstone models. PhD book. Department of Chem Material Eng, University of Alberta

Hogg, R. 1999. Breakage mechanisms and mill performance in ultrafine grinding. Powder Technology. 105(1):135-140.

Huang, H., Detournay, E. & Bellier, B. 1999. Discrete element modelling of rock cutting. Rock Mechanics for Industry. 1(1):123-130.

Hudson, J.A. & Harrison, J.P. 2000. Engineering rock mechanics: an introduction to the principles. Elsevier.

Inglis, C.E. 1913. Stresses in a plate due to the presence of cracks and sharp corners. Trans Inst Naval Archit. 55:219-241.

International Society for Rock Mechanics. 2007. The complete ISRM suggested methods for rock characterisation, testing and monitoring: 1974-2006. International Soc. for Rock Mechanics, Commission on Testing Methods.

International Society for Rock Mechanics, 1979. ISRM—suggested methods for determining the uniaxial compressive strength and deformability of rock materials.

Irwin, G.R. 1947. Fracture dynamics. Fracturing of Metals.

Irwin, G.R. 1956. Onset of Fast Crack Propagation in High Strength Steel and Aluminum Alloys.

Itasca. 1999. PFC3D Manual (1st edn). Itasca Consulting Group, Inc., Minneapolis, Minesota.

Ivars, D.M., Pierce, M.E., Darcel, C., Reyes-Montes, J., Potyondy, D.O., Young, R.P. & Cundall, P.A. 2011. The synthetic rock mass approach for jointed rock mass modelling. International Journal of Rock Mechanics and Mining Sciences. 48(2):219-244.

Jaeger, J.C., Cook, N.G. & Zimmerman, R. 2009. Fundamentals of rock mechanics. John Wiley & Sons.

Jiménez-Herrera, N., Barrios, G.K. & Tavares, L.M. 2017. Comparison of breakage models in DEM in simulating impact on particle beds. Advanced Powder Technology.

Jing, L. 2003. A review of techniques, advances and outstanding issues in numerical modelling for rock mechanics and rock engineering. International Journal of Rock Mechanics and Mining Sciences. 40(3):283-353.

Johnston, I.W. & Choi, S.K. 1986. A synthetic soft rock for laboratory model studies. Geotechnique. 36(2):251-263.

Kanninen, M.F. & Popelar, C.H. 1985. Advanced fracture mechanics. Oxford University Press.

Kapur, P.C., Pande, D. & Fuerstenau, D.W. 1997. Analysis of single-particle breakage by impact

King, R.P. 2012. Modeling and simulation of mineral processing systems. Elsevier.

King, R.P. & Bourgeois, F. 1993. Measurement of fracture energy during single-particle fracture. Minerals Engineering. 6(4):353-367.

Kojovic, T., Shi, F., Larbi-Bram, S. & Manlapig, E. 2008. Julius Kruttschnitt Rotary Breakage Tester (JKRBT): Any ore, any mine. MetPlant 2008. AusIMM. 91-103.

Krogh, S.R. 1980. Crushing characteristics. Powder Technology. 27(2):171-181

Kuhn, C. and Müller, R., 2010. A continuum phase field model for fracture. Engineering Fracture Mechanics, 77(18), pp.3625-3634.

Labuz, J.F. & Zang, A. 2012. Mohr–Coulomb failure criterion. Rock Mechanics and Rock Engineering. 45(6):975-979.

Lee, C., Jeon, S. & Lee, Y. 2011. Simulation of mechanical and thermal behavior of synthetic rock around an opening using bonded particle model. Proceedings of 2nd International Symposium on Computational Geomechanics, Cavtat-Dubrovnik, Croatia. 870-879.

Le Pham, T.T. 2011. Modelling of Breakage Phenomena in Rocks: Mathematical and Numerical Modelling Concepts and Analysis-Modelling of Elastic Property, Mechanics and Physics Problems, Fracture Patterns and Mechanisms Using Discrete Element Methods, and Applying Theory to M.

Liu, H.Y., Kou, S.Q., Lindqvist, P. & Tang, C.A. 2007. Numerical modelling of the heterogeneous rock fracture process using various test techniques. Rock Mechanics and Rock Engineering. 40(2):107-144.

Johannes, M., Quist, J. Evertsson, M. 2016. Bonded particle model calibration using design of experiments and multi-objective optimisation. MEI 10th International Comminution Symposium (Comminution '16).

Mandal, T.K., Nguyen, V.P. and Wu, J.Y., 2019. Length scale and mesh bias sensitivity of phase-field models for brittle and cohesive fracture. Engineering Fracture Mechanics, 217, p.106532.

Matuttis, H. & Chen, J. 2014. Understanding the discrete element method: simulation of non-spherical particles for granular and multi-body systems. John Wiley & Sons.

McLaren, C.H. & De Villiers, J.P. 1982. The platinum-group chemistry and mineralogy of the UG-2 chromitite layer of the Bushveld Complex. Economic Geology. 77(6):1348-1366.

Mei, C., Fang, Q., Luo, H., Yin, J. & Fu, X. 2017. A synthetic material to simulate soft rocks and its applications for model studies of socketed piles. Advances in Materials Science and Engineering.

Mishra, B.K. & Thornton, C. 2001. Impact breakage of particle agglomerates. International Journal of Mineral Processing. 61(4):225-239.

Montgomery, D.C. 2017. Design and analysis of experiments. John Wiley & sons.

Mora, P. & Place, D. 1993. A lattice solid model for the nonlinear dynamics of earthquakes. International Journal of Modern Physics C. 4(06):1059-1074.

Morrison, R.D., Shi, F. & Whyte, R. 2007. Modelling of incremental rock breakage by impact–for use in DEM models. Minerals Engineering. 20(3):303-309.

Mwale, A.N., Mainza, A.N., Bepswa, P.A., Frausto J.J.F., Ballantyne, G., Cruz, R., Gomez, S. 2018. Fitting the fine screen model into industrial plant data. XXIX International Mineral Processing Congress, Moscow, Russia.

Napier-Munn, T.J., Morrell, S., Morrison, R.D. & Kojovic, T. Mineral Comminution Circuits–Their Operation and Optimization, (1996). JKMRC Monograph Series in Mining and Mineral Processing. 2.

Nappier Munn, A.L., Morrell, S. & Kojovic, T. 1999. Mineral comminution circuits (Their operation and optimization). The University of Queensland: JKMRC Monograph Series in Mining and Minera L Processing. 2.

Narayanan S.S. & Whiten, W.J. 1988. Determination of comminution characteristics from single-particle breakage tests and its application to ball-mill scale-up. Transactions of the Institution of Mining and Metallurgy Section C-Mineral Processing and Extractive Metallurgy. 97:C124.

Nesetova, V. & Lajtai, E.Z. 1973. Fracture from compressive stress concentrations around elastic flaws. International Journal of Rock Mechanics and Mining Sciences & Geomechanics Abstracts. Elsevier. 265-284.

Nong, X. & Towhata, I. 2017. Investigation of mechanical properties of soft rock due to laboratory reproduction of physical weathering process. Soils and Foundations. 57(2):267-276.

Norgate, T. & Haque, N. 2010. Energy and greenhouse gas impacts of mining and mineral processing operations. Journal of Cleaner Production. 18(3):266-274.

Oladele T.P., Bbosa L.B., & Weatherley D.K. 2019. Evaluation of the bonded particle model for simulating rock breakage under impact loading. Minerals Engineering (Under review)

Orowan, E. 1949. Fracture and strength of solids. Reports on Progress in Physics. 12(1):185.

Pauw, O.G. & Maré, M.S. 1988. The determination of optimum impact-breakage routes for an ore. Powder Technology. 54(1):3-13

Petruk, W. 2000. Applied mineralogy in the mining industry. Elsevier.

Pharr, G., Oliver, W. & Brotzen, F. 1992. On the generality of the relationship among contact stiffness, contact area, and elastic modulus during indentation. Journal of Materials Research. 7(03):613-617.

Pickering, R. & Ebner, B. 2002. Hard rock cutting and the development of a continuous mining machine for narrow platinum reefs. Journal-South African Institute of Mining and Metallurgy. 102(1):19-24.

Place, D.G. & Mora, P.R. 2001. A random lattice solid model for simulation of fault zone dynamics and fracture process. Bifurcation and Localisation Theory for Soils and Rocks' 99. 1.

Potapov, A.V. & Campbell, C.S. 1994. Computer simulation of impact-induced particle breakage. Powder Technology. 81(3):207-216.

Potapov, A.V. & Campbell, C.S. 2001. Parametric dependence of particle breakage mechanisms. Powder Technology. 120(3):164-174.

Potapov, A., Herbst, J.H., Song, M. & Pate, W. 2007. A DEM-PBM fast breakage model for simulation of comminution processes. Proceedings of the MEI Discrete Element Methods (DEM) 07 Conference.

Potyondy, D.O. & Cundall, P.A. 2004. A bonded-particle model for rock. International Journal of Rock Mechanics and Mining Sciences. 41(8):1329-1364.

Potyondy, D.O., Cundall, P.A. & Lee, C.A. 1996. Modelling rock using bonded assemblies of circular particles. 2nd North American Rock Mechanics Symposium. American Rock Mechanics Association.

Potyondy, D.O. 2015. The bonded-particle model as a tool for rock mechanics research and application: current trends and future directions. Geosystem Engineering. 18(1):1-28.

Powell, M.S. & Weerasekara, N. 2010. Building the unified comminution model. XXV International Mineral Processing Congress Brisbane Australia. 1133-1142.

Quist, J. & Evertsson, C.M. 2016. Cone crusher modelling and simulation using DEM. Minerals Engineering. 85:92-105.

Radziszewski, P. 2013. Energy recovery potential in comminution processes. Minerals Engineering. 46:83-88.

Radziszewski, P. & Laplante, A. 2006. Exploring the effect of friction losses on drop weight breakage results. AMIRA Project P9N Mineral Processing 5th P9N Technical Report. 42–49

Rao, G.N. & Murthy, C.R. 2001. Dual role of microcracks: toughening and degradation. Canadian Geotechnical Journal. 38(2):427-440.

Reynolds, G.K., Fu, J.S., Cheong, Y.S., Hounslow, M.J. & Salman, A.D. 2005. Breakage in granulation: a review. Chemical Engineering Science. 60(14):3969-3992.

Rocco, C., Guinea, G., Planas, J. & Elices, M. 1999. Size effect and boundary conditions in the Brazilian test: theoretical analysis. Materials and Structures. 32(6):437-444

Rule, C. & Schouwstra, R.P. 2012. Process mineralogy delivering significant value at Anglo Platinum concentrator operations. Proceedings of the 10th International Congress for Applied Mineralogy (ICAM). Springer. 613-621.

Salman, A.D., Ghadiri, M. & Hounslow, M. 2007. Particle breakage. Elsevier.

Salman, A., Reynolds, G. & Hounslow, M. 2003. Particle impact breakage in particulate processing. KONA Powder and Particle Journal. 21(0):88-99.

Schön, J.H. 2015. Physical properties of rocks: Fundamentals and principles of petrophysics. Elsevier.

Schouwstra, R.P., Kinloch, E.D. & Lee, C.A. 2000. A short geological review of the Bushveld Complex. Platinum Metals Review. 44(1):33-39.

Saeidi, F., Yahyaei, M., Powell, M. and Tavares, L.M., 2017. Investigating the effect of applied strain rate in a single breakage event. Minerals Engineering, 100, pp.211-222.

Sharon, E., Gross, S.P. & Fineberg, J. 1995. Local crack branching as a mechanism for instability in dynamic fracture. Physical Review Letters. 74(25):5096.

Shi, F. & Kojovic, T. 2007. Validation of a model for impact breakage incorporating particle size effect. International Journal of Mineral Processing. 82(3):156-163.

Sjöberg, J. 1997. Estimating rock mass strength using the Hoek–Brown failure criterion and rock mass classification—a review and application to the Aznalcollar open pit. Division of Rock Mechanics, Department of Civil and Mining Engineering, Lulea University of Technology. 154.

Smith, S.E., MacLaughlin, M.M., Adams, S.L., Wartman, J., Applegate, K., Gibson, M., Arnold, L. & Keefer, D. 2014. Interface Properties of Synthetic Rock Specimens: Experimental and Numerical Investigation. Geotechnical Testing Journal. 37(5):840-858.

Song, B. & Chen, W. 2005. Split Hopkinson pressure bar techniques for characterising soft materials. Latin American Journal of Solids and Structures. 2(2):113-152.

Squelch, A. 2018. 3D printing rocks for geo-educational, technical, and hobbyist pursuits. Geosphere. 14(1):360-366.

Tavares, L.M. 1999. Energy absorbed in breakage of single particles in drop weight testing. Minerals Engineering. 12(1):43-50.

Tavares, L.M.M. 1998. Microscale investigation of particle breakage applied to the study of thermal and mechanical predamage. PhD book. Department of Metallurgical Engineering, The University of Utah

Tavares, L.M., André, F.P., Potapov, A. and Maliska Jr, C., 2020. Adapting a breakage model to discrete elements using polyhedral particles. Powder Technology, 362, pp.208-220.

Tavares, L.M. & King, R.P. 1998. Single-particle fracture under impact loading. International Journal of Mineral Processing. 54(1):1-28.

Tavares, L.M. & King, R.P. 2004. Measurement of the load–deformation response from impact-breakage of particles. International Journal of Mineral Processing. 74: S277.

Tavares, L.M. 2007. Breakage of single particles: quasi-static. Handbook of Powder Technology. 12:3-68.

Thornton, C., Ciomocos, M.T. & Adams, M.J. 1999. Numerical simulations of agglomerate impact breakage. Powder Technology. 105(1):74-82.

Tromans, D. 2008. Mineral comminution: energy efficiency considerations. Minerals Engineering. 21(8):613-620.

Ulusay, R. 2014. The ISRM suggested methods for rock characterisation, testing and monitoring: 2007-2014. Springer.

Unland, G. & Al-Khasawneh, Y. 2009. The influence of particle shape on parameters of impact crushing. Minerals Engineering. 22(3):220-228.

Unland, G. & Szczelina, P. 2004. Coarse crushing of brittle rocks by compression. International Journal of Mineral Processing. 74: S217.

Verret, F.O., Chiasson, G. & McKen-SGS, A. 2011. SAG mill testing-an overview of the test procedures available to characterise ore grindability. International Autogenous and Semiautogenous Grinding and High-Pressure Grinding Roll Technology.

Vogel, L. & Peukert, W. 2004. Determination of material properties relevant to grinding by practicable labscale milling tests. International Journal of Mineral Processing. 74: S329–S338.

Wang, M. & Cao, P. 2017. Calibrating the micromechanical parameters of the PFC2D (3D) models using the improved simulated annealing algorithm. Mathematical Problems in Engineering. 2017.

Wang, Y., Abe, S., Latham, S. & Mora, P. 2006. Implementation of particle-scale rotation in the 3-D lattice solid model. Pure and Applied Geophysics. 163(9):1769-1785.

Wang, Y. & Mora, P. 2009. The ESYS_particle: a new 3-D discrete element model with single particle rotation. In Advances in Geocomputing. Springer. 183-228.

Wang, Y. & Tonon, F. 2010. Calibration of a discrete element model for intact rock up to its peak strength. International Journal for Numerical and Analytical Methods in Geomechanics. 34(5):447-469.

Wang, Y., Yin, X., Ke, F., Xia, M. & Peng, K. 2000. Numerical simulation of rock failure and earthquake process on mesoscopic scale. Pure and Applied Geophysics. 157(11-12):1905-1928.

Weatherley, D. 2013. Numerical investigations relating the lithology and mechanical properties of rock. 13th European Symposium on Comminution and Classification. Braunschweig, Sierke Verlag; 2013:221-224.

Weatherley, D. & Ayton, T. 2012. Numerical investigations on the role of micro-cracks in determining the compressive and tensile strength of rocks. EGU General Assembly Conference Abstracts. 8294.

Weatherley, D., Boros, V. & Hancock, W. 2011. ESyS-particle tutorial and user's guide Version 2.1. Earth Systems Science Computational Centre, the University of Queensland.

Weerasekara, N.S., Powell, M.S., Cleary, P.W., Tavares, L.M., Evertsson, M., Morrison, R.D., Quist, J. & Carvalho, R.M. 2013. The contribution of DEM to the science of comminution. Powder Technology. 248:3-24.

Weichert, R. & Herbst, J.A. 1986. An ultra-fast load cell device for measuring particle breakage. 1st World Congress of Particle Technology. Nürnberg.

William, P. 2014. Numerical investigations on rock breakage in minerals processing using supercomputers. Master's book. The University of Queensland, Australia.

Wills, B.A. & Napier-Munn, T. 2015. Wills' mineral processing technology: an introduction to the practical aspects of ore treatment and mineral recovery. Butterworth-Heinemann.

Wong, R.H. & Chau, K.T. 1998. Crack coalescence in a rock-like material containing two cracks. International Journal of Rock Mechanics and Mining Sciences. 35(2):147-164.

Xiao, Z. & Laplante, A.R. 2004. Characterising and recovering the platinum group minerals––a review. Minerals Engineering. 17(9-10):961-979.

Yahyaei, M., Hilden, M., Shi, F., Liu, L.X., Ballantyne, G. & Palaniandy, S. 2016. Comminution. In Production, Handling and Characterization of Particulate Materials. Springer. 157-19.

Yoon, J. 2007. Application of experimental design and optimization to PFC model calibration in uniaxial compression simulation. International Journal of Rock Mechanics and Mining Sciences. 44(6):871-889.

Zhai, S., Zhan, J., Ba, Y., Chen, J., Li, Y. & Li, Z. 2019. PFC model parameter calibration using uniform experimental design and a deep learning network. IOP Conference Series: Earth and Environmental Science. IOP Publishing. 032062.

Zhang, Y. (2010). Probabilistic calibration of a discrete particle model. Doctoral book. USA: Texas A&M University.

Zhou, T. & Zhu, J.B. 2018. Identification of a suitable 3D printing material for mimicking brittle and hard rocks and its brittleness enhancements. Rock Mechanics and Rock Engineering. 51(3):765-777.